T0082531

The LITTLE BOOK *Of* MANAGING UNCERTAINTY

LITTLE BOOKS

The LITTLE BOOK *of* ARTIFICIAL INELLIGENCE

The LITTLE BOOK *of* SERVICE MANAGEMENT

The LITTLE BOOK *of* CYBERSECURITY

The LITTLE BOOK *of* CLOUD COMPUTING

The LITTLE BOOK *of* MANAGING UNCERTAINTY

The LITTLE BOOK *Of* MANAGING UNCERTAINTY

HARRY KATZAN JR.

THE LITTLE BOOK OF MANAGING UNCERTAINTY

iUniverse books may be ordered through booksellers or by contacting:

iUniverse
1663 Liberty Drive
Bloomington, IN 47403
www.iuniverse.com
844-349-9409

ISBN: 978-1-6632-4399-7 (sc)
ISBN: 978-1-6632-4400-0 (hc)
ISBN: 978-1-6632-4401-7 (e)

Library of Congress Control Number: 2022915156

Print information available on the last page.

iUniverse rev. date: 08/18/2022

To Margaret, Kathy, and Karen

PREFACE

This little book is an easy-to-read collection of chapters written by the author on the subject of Managing Uncertainty. Some chapters have been written to suit a general audience, and others have been prepared for a select class of readers with specific problems. There is some mathematics involved in some topics, but it is very general in nature. No specific math background is required at all.

The topics are intended to be read separately resulting in a minimal amount of definitional material being repeated throughout the book. The reader is able to comfortably read the entries on a topic of interest and disregard the remainder. The chapters are related, but each has a unique focus.

Uncertainty, covered later in the book, deals mainly in situations where the problem definition is not known or the result is not known or understood. Here are a couple of exam. In the first case, a good engineering student going into his or her senior with all A's and no B's, C's, or anything lower. The student has two options. His advisor tells him he can get a Master's degree in engineering in one year. On the other hand, he can go into industry with a good paying job. The probably that he would achieve his Master's is practically 1.0, given his record. The student may not be so fortunate with the job. He might not like his work assignment, the company atmosphere, his manager, or the living environment. A lower probably is expected, ranging from 0.5 to 0.75. If the student is risk averse, he should take the Master's. Otherwise the job might be appropriate.

Another example might involve a soldier in a desert situation of some sort. The solder is manning an artillery device in the desert and spots a cloud of dust in the distance. The question is: Is it a bunch of wild camels, or an enemy force moving across the desert. Before firing or not firing, the solder asks his sergeant who responds that the cloud is causes by a bunch of camels, a bunch of desert residents, or an enemy force with probabilities (0.70, 0.25, 0.05) respectively. Not satisfied, he asks his commanding officer who responds with (0.90, 0.10, 0.00). The soldier has a decision to make and askes a CIA officer who accompanies the group. Using Dempster's Rule of combination, the officer comes up with (0.72, 0.28, 0.00), yielding the following assessment:

Camels – 0.72 (72%)
Residents – 0.28 (28%)
Enemy force – 0.00 (0%)

Arthur Dempster is a well-known Harvard professor who developed Dempster's Rule of combination as a Harvard.

As simple as it seems, calculations, such as these, is the basis of understanding uncertainty. This book contains introductory material on understanding uncertainty, relevant mathematical background, and useful examples.

The reader can use the material in the book in several ways. The first part is an easy to read description of uncertainty with common everyday examples. Included is a description of the mathematics of uncertainty using Dempster-Shafer theory. It requires no specific background in math but involves careful study. The third part gives six useful applications of the underlying concepts.

The appendix gives an introductory section on Dempster-Shafer theory. It should be read first.

Enjoy reading the book. You will be glad you did.

Harry Katzan, Jr

CONTENTS

Part One: Introduction to Uncertainty

1 Managing Uncertainty: A Pragmatic Approach 1

2 Uncertainty Information Blocks 15

3 Dempster-Shafer Theory of Evidence 25

4 The New Connectionism 35

Part Two: Applications of Uncertainty

5 Identity Analytics and Belief Structures 43

6 Evolutionary Dynamics of Service Provisioning 61

7 Categorical Analytics Based on Consensus Theory 73

8 Toward a Unified Ontology of Trusted Identity in Cyberspace 87

9 Product Analytics Based on Demographic Democratization 113

10 Structural Analytics for Decision Making under Uncertainty 131

Appendix: A Brief Introduction 145

Index 151

About The Author 155

Part One

INTRODUCTION TO UNCERTAINTY

1

MANAGING UNCERTAINTY: A PRAGMATIC APPROACH

INTRODUCTION

Managers, scientists, and analysts in today's world face a common problem. In fact it is a situation unique in the history of science and commence. Never before have we had the unparalleled capability and capacity to collect and process information as is currently available with modern high-technology computer systems. We have arrived at a point in time – a reference point, so to speak – where one of the key elements in science and business is what to do with the available information, and there would seem to be enough of it around to turn a minor consideration into a major concern.

Information Domain

More specifically, the information problem has two facets:

- *What* information should be collected (*specificity*).
- *How* should the information be utilized in order to give decision makers the needed values for key decision variables (*functionality*).

Our concern is with specificity and functionality from the point of view of information. In fact, the major objective is to take a fresh approach to information where distinct elements can be viewed as though they form a system with various pieces interacting to a high degree.

The discussion of specificity involves how to select information that leads to more definitive knowledge about a problem domain. The discussion of functionality involves how to use computer programs to generate results that are appropriate to the needs of a particular decision maker for a more general audience. It is clearly recognized that this process of information distillation has both specificity and functionality components. The opposite is also true for the reverse direction. However, the straightforward notion of information flow is simply not sophisticated and does not constitute a purposeful system. Information should be interpreted and transformed as it flows through a system and the process of transformation should be under the control of relations and mappings developed by domain specialists.

Indicators

In the real world, information exists in a wide variety of important forms. Written material and electronically recorded data are the primary modes of storage, but information is also inherent in conventions, procedures, designs, social customs, and so forth. In fact, whenever a decision has been taken, one can informer information implicit to the alternatives that were selected and to those rejected. There is a wealth of information, in evolutionary processes vis-á-vis the forces that control evolution - regardless if the process is biological, social, or organizational. Another way of saying this is that there is more information (and less uncertainty) in systems of rules than on the specific symbols on which the rules operate.

It is useful to put a noose around a class of information and anchor it to a model representing the area under analysis. In this context, it is a customary to employ indicators as a measure of conditions within a system. Economists rely on leading indicators as a measure of conditions at a point time. Similarly, meteorologists use indicators, such as temperature, dew-point, and barometric pressure – to name only a few.

An *indicator* is a variable that is instantiated by assigning it a value. As a simple example in meteorology, assume the dew point is 75 degrees. Thus the instantiated variable is "dew point." By itself, the value of a variable is a small amount of information. But in the context of a problem space, it implies a lot (see Information Block # 1); it implies, in this case, that the weather is unusually muggy and uncomfortable. There is some uncertainty involved because temperature and other conditions also play a part in the comfort level of the weather.

PROBLEM DOMAIN

Uncertainty exists when there is some question as to the true value of a proposition. The set of possibilities is represented by a frame of discernment (see Information Block # 3), and relevant evidence is inherent in the indicators that supply information about the problem domain.

Input Space

The *input space* reflects information represented by the values of an indicator variable and a problem domain specified by one or more frames of discernment.

The process involves gleaning information from indicator variables and subsequent combining this information to form belief structures. Indicator variable represent world conditions that are translated into belief sets through input mappings. The approach is clearly Bayesian (see Information Block #2).

The input space, in a real sense, is used to encapsulate a decision situation. Later, an output space is covered that serves an analogous purpose for the solution.

Input Mapping

An *input mapping* is established by a domain specialist and exists as a straightforward one-to-one correspondence between the value of an indicator variable and a focal set over a predefined frame of discernment. As a simple medical example, assume a frame of discernment in the input problem space, called in this instance the *diagnosis*, and delineated as:

diagnosis = {healthy, tests, sick}

In the environment, there are three indicator variables: *lump, blood,* and *color,* with the following possible values:

lump: {none, small, large}
blood: {string, of, weak}
color: {pale, normal}

Clearly, each indicator variable can be associated with only one of its possible values at a given time, so a sample scenario could be:

lump = small
blood = weak
color = normal

An input mapping is a relationship between a scenario and a subset of the frame of discernment. Here is a set of input mappings:

Variable	*Value*	*Focal Set*
lump	none	{{healthy}, 0.5}
lump	small	{{healthy, tests} 0.6}

lump	large	{{tests, sick,}, 0.9}
blood	strong	{{healthy}, 0.8}
blood	ok	{{healthy, tests}, 0.7}
blood	weak	{{sick}, 0.0.9}
color	pale	{{tests, sick}, 0.7}
color	normal	{{healthy}, 0.5}

Thus, the proper input mappings applied to the sample scenario yield the following evidence:

Indicator	*Value*	*Simple Support Unit*
lump	small	{{{healthy, tests}, 0.6}, {{health, test, sick}, 0.4}}
blood	weak	{{{sick}, 0.9}, {{health, test, sick}, 0.1}}
color	normal	{{{healthy}, 0.5}, {{health, test, sick}, 0.5}}

So far, the indicator variables and their associated input mappings reflect the same frame of discernment in the input space. Therefore, it is feasible to combine this evidence using Dempster's rule of combination.

The input space may incorporate more than one frame of discernment; however, only evidence defined over the same frame may be combined. A given indicator variable may be associated with one and only one frame, but the is an operational expedient. An indicator that maps to two frames does not violate the independence rule since that evidence in never combines with Dempster's rule of combination.

FUSING OF EVIDENCE

The *fusing of evidence* refers to the narrowing process inherent in combining evidence. Simple support functions are combined to form separable support functions. Actually, it is a cascading process. At first, two simple support functions are combined to form a separable support function, which is then combined with another simple support function, and so forth.

In the previous example, fusing *lump* = small with *blood* = weak yields:

> {{{sick}, 0.36}, {{healthy, tests} 0.06}, {{healthy, tests, sick}, 0.04}, {null, 0.54}}

Which is then normalized to:

> {{{sick}, 0.78}, {{healthy, tests} 0.13}, {{healthy, tests, sick}, 0.09}}

By dividing the non-null entries by (1-*bpa*) of the null entry, which in this case is 0.46. Note that the set {healthy, test, sick} is the frame of the discernment.

To continue, fusing *color* = normal yields:

> {{{sick}, 0.39}, {{healthy, 0.11}, {{healthy, tests}, 0.07}, {{healthy, test, sick}, 0.04}, {null, 0.39}}

which is normalized to:

> {{{sick}, 0.64}, {{healthy, 0.18}, {{healthy, tests}, 0.11}, {{healthy, test, sick}, 0.07}}

It is interesting from this elementary example how a value of "normal" for the indicator variable *color* served to diffuse the *bpa* for the propositions {sick}, which moves from o.78 to 0.64.

It is now clear why a domain specialist would establish indicator variables and define input mappings. The end user, of course, would only perform the following tasks:

- Instantiate indicator variables.
- Activate input mappings.
- Fuse evidence (often referred to as "fusing the frame").

The domain specialist can do more. It is expected that a specialist would know what the fused evidence means and what to do with it.

SOLUTION DOMAIN

The solution domain incorporates a rule-based component that gives the flavor of being an expert advisor. The methodology employs expert-systems technology and adheres to the advisor paradigm. The problem domain is regarded as the input space, and the solution domain is known as the output space.

Output Space

The *output space* reflects what action should be taken in response to certain values in the input space. For example, knowledge that the belief accorded to the proposition {healthy} is 0.11 does not *a priori* help in determining what to do with that knowledge. The first consideration is to change the frame of discernment. Some options for the previous example would be:

Proposition	*Meaning*
{office}	Go to the office for routine tests.
{office, hospital}	Go to the office or hospital for extensive tests.
{hospital}	Check in to the hospital. You're sick.
{home}	Stay home. You're healthy.

Accordingly, the frame of discernment is:

$$\Theta = \{\text{office, hospital, home}\}$$

and the set of subsets of Θ is:

$$2^{\Theta} = \{\{\text{office}\}, \{\text{hospital}\}. \{\text{home}\}, \{\text{office, hospital}\}, \{\text{office, home}\}, \{\text{hospital, home}\}, \{\text{office., hospital, home}\}\}$$

Clearly, every element of 2^Θ may not be meaningful need not be used.

The output space incorporates decision entities, known as DSS objects, and decision scripts that are covered later. DSS objects, where DSS stands for Decision Support System, are the basis on which decisions are made and decision scripts give possible actions that could be taken.

Compatibility Relations

A *compatibility relation* is a correspondence between focal sets in two spaces, In this context, the concept is used to map entries from the input space to the output space. Entries in the input space are referred to as the *antecedent* m and the output space are known as the *consequent.*

An example of a compatibility relation is:

({healthy, tests}, 0.6) -> ({home. Office}, 0.7)

It should be interpreted as follows: if the *bpa* of the focal set {healthy, test} in the antecedent exceeds the threshold of 0.6, replace it in the consequent with the focal set {home, office} and compute its *bpa* (i.e., the new *bpa*) as the *bpa* from the antecedent times the certainty value of 0.7. The result is summarized in the following list:

Antecedent	*Consequent*
... {{healthy, tests}, 0.8} ...	... {{home, office}, 0.56} ...

Thus, the form of a compatibility relation is:

(antecedent, threshold) → (consequent, CV)

Where CV is the Certainty Value of the relation.

Compatibility relations are commonly known as *rules* that are used according to the following procedure:

- Each entry in the antecedent is matched against the set of rules.
- If the focal sets match and the *bpa* of the focal set from the antecedent exceeds the threshold, then that rule is selected for firing.
- A rule is fired by selecting the consequent focal set and assigning a *bpa* of the antecedent focal set multiplied by the certainty value.

There is a major difference, however, between focal sets in the input space and focal sets in the output space. In the input space, all focal sets over the same frame are part of a simple or separable support unit. In the output space, focal sets are fragments of support units, possibly from different frames of discernment. It follows that focal sets must be fused in the output space as well as in the input space.

A domain specialist would also establish compatibility relations.

Output Mapping

Output mapping provides a means of interpreting the result of the mapping operation from the input to the output spaces. With supporting, as well as conflicting evidence, there is normally a certain amount of noise information that can and should be filtered out. Interpretation, DSS objects, and decision scripts are important for the filtering process.

Interpretation

Interpretation is the process of fusing evidence in the output space and subsequently applying a threshold system that effectively masks out unneeded details.

Using the frame of discernment

Θ = {office, hospital, home}

and the previous example, an element in the input space is mapped to the output space; the process generates the following focal set:

{{home, office}, 0.56}

The first task performed by interpretation is to transform the result of each mapping into a simple support set; in the example, it is

{{home, office}, 0.56}, {{office, hospital, home}, 0.44}}

After all of the compatibility relations have been "fired," simple support sets over the same frame are fused using Dempster's rule of combination. The result is a separable support set for each frame in the output space. The threshold system comes next.

In practical applications, the process of input mapping, fusing, compatibility mapping, and then more fusing generates noise data consisting of focal sets with very small *bpa*s. Thus, it is prudent to provide a "cover" for the computed results by filtering out this unneeded noise.

It follows that an *output set,* consisting of a focal set - i.e., a focal set in the output set – and a threshold is specified. Each focal set with a *bpa* that exceeds its declared threshold becomes a DSS object. This is the first part of a two-stage filtering system. The second part involves the generation of *What To Do*'s called decision scripts.

Using a two-stage filtering system has certain advantages. First it allows the cover to be readily changed to more closely track the changing needs of an evolving decision environment. Secondly, it permits the system to be tuned by adjusting the noise level.

DSS Objects

A *DSS object*, where DSS stands for Decision Support System, is a belief structure that is relevant to the needs of a decision maker. More specifically, a DSS object is a focal set and its *bpa*. One or more DSS objects may be generated from a separable support set over a given frame of discernment in the output space.

A DSS object is a form of information. It is evidence, which has not been used up, so to speak, in the compatibility mapping. On the other hand, a DSS object is not a "What To Do." This is where the second part of the two-stage filtering system comes in.

In a real sense, a DSS object is a candidate for further study. The DSS object has passed the relevancy test and noise information has been eliminated. There are other considerations.

Some DSS objects make sense *only* if the *bpa* is relatively high. Moreover, a given DSS object may only pertain to a specific kind of decision or a particular brand of decision maker. Furthermore, two decision makers may look at the same DSS object but at differing threshold levels.

Action Items

The *What To Do*, referred to earlier, is simply an action item. Decision makers apparently prefer action items to a collection of number and this approach supports that philosophy.

An *action item* is an activity or amenity that has meaning in a decision-making situation, such as one of the following:

- The generation of an order, comment, or assertion.
- The initiation of the printing of a report or the display of information.
- The energizing of a physical device, such as a light or switch.
- The invocation of the execution of a coputer program.
- The progression of a system to another state.

In other words, an action item is something that has meaning to a decision maker in the context of the output space.

In the present application of Dempster-Shafer theory, an action item is always associated with a DSS object and a threshold. If the *bpa* of the focal set is equal to or exceeds the given threshold value, then the action item is selected.

A given action item may be defined over one and only one DSS object. However, several different actions may be defined over the same DSS object. If it is necessary to relate the same action to more than one DSS object, then that action only need be defined the requisite number of times.

Decision Scripts

In the System of managing uncertainty, covered herein, the action item implemented is a decision script. A *decision script* takes the following information:

- Output set name.
- Threshold value.
- Action item specifying a message.

The output set gives the frame name and the focal set that are relevant to the associated DSS object.

SUMMARY

In today's world, and unparalleled capacity exists for collecting and processing information. Thus, the biggest problems facing managers, scientists, and analysts are *what* information should be collected and *how* it should be utilized. This problem should be faced from the view of information.

One of the most useful forms of information is an indicator, which encapsulates a class of date and anchors it to a paradigm representing an area under analysis. More specifically, and indicator is the value of a variable, such as the gross national product in economics or the dew point in meteorology.

Indicators mean different things in different problem situations – even though a certain amount of interpretation is implicit in the paradigm to which they are associated. To be really useful, however, an indicator must be transformed to an input "problem" space through an input mapping.

An input space represents a decision situation. Evidence is combined in the input space and then mapped to an output space through compatibility relations, Decision support system objects (DSS objects) are interpreted in the output space to provide action items for decision makers.

REFERENCES

Gordon, J. and E. Shortliffe, "A Method of Management Evidential Reasoning in a Hierarchical Hypothesis," *Artificial Intelligence 26,* 323-357.

Kanak, L. and j. Lemer, *Uncertainty in Artificial Intelligence,* Amsterdam, 1986.

Katzan, H., "A Prototype Shell of the Modeling of Uncertainty," University of Fribourg # 133, 1987.

Katzan, H., *Managing Uncertainty,* New York, 1992.

Shafer, G., *A Mathematical Theory of Evidence, Princeton,* 1976.

END OF CHAPTER ONE

2

UNCERTAINTY INFORMATION BLOCKS

INTRODUCTION

This chapter includes three information blocks listed as follows:

The Advisor Model #1
Dealing With Uncertainty #2
Practical Considerations #3

The blocks should provide definitive information in addition to background material

THE ADVISOR MODEL – Block #1

An Advanced Decision-System (ADSS) is characterized by a data base of facts and a processing facility for exploiting the facts in a purposeful manner. The word "exploiting,"

Used in this context, is intended to denote that the computing mechanism takes good advantage of the available information so that the output of the system is considerably more that when would ordinarily be implied strictly by the input.

The Proverbial Wine Advisor

One of the simplest examples of a ADSS is the well know wine advisor. The user enters a brief description of an entrée, such as chicken with cream sauce, and the system yields a small wine selection, such as California Chardonnay, Swiss Fendant, and French Montrachet – depending of course upon the current contents of the wine cellar. Naturally, there is more to the story. The entrée has certain characteristics that are determined by a data base search. The characteristics are then used by an inference mechanism during a "run" of the system to generate the attributes of a suitable wine, such as color, taste (i.e., degree of dryness), and strength (i.e., light, heavy, etc.). In a second data base operation, the wine attributes are mapped into a "variety" of wine. Finally, a selection is made from the data base that represents the contents of the wine cellar.

Similar advisors have been constructed in the financial community and in the nuclear industry – to name only a few. In an investment advisor, for example, and investor's profile is mapped into a well-defined, but much larger, set of investment characteristics using a data base operation. Again, an inference process is used; in the case, it generates a portfolio structure that serves as input to a final data base search operation yielding a set of investment options. A typical portfolio structure might be: 60% equities, 35% bonds, and 5% gold. The output set would be a small selection of instruments in selected currencies.

In the nuclear industry, a collection of physical events occurring in an emergency situation are effectively translated by an ADSS into a set of instructions to an operator. As before, the processing sequence includes an initial data base search operation, an inference process, and them a final data base selection to generate the needed instructions. The three examples typify the "advisor model."

Description of the Model

The *advisor model* is characterized by the following set of actions occurring in essentially the order given:

- A relatively small set of input values ae entered into the system via an end user interaction or as the result of a physical event.
- The input values are expanded into the particulars for a certain problem domain via a data base search operation.
- The particulars are mapped into a set of descriptors via an inference process that can take the form of a set of it-then rules and a logical engine, a neural network, a pattern matching operation, a mathematical function, or even possibly a logic program.
- The descriptors are used as key values in a subsequent data base search operation to yield the results for that problem.

Clearly, the salient properties of the advisor model are the initial data base operation, the inference (or mapping) process, and the final data base search. (The two data base operations are obviously to different structures.} Each of these components can be replicated, as required by a particular problem and its problem solver, but the general structure of the model is evident without necessarily going into more detail.

Uncertainty

Another consideration that complicates the picture is the existence of uncertainty in most areas dealing with human affairs. Uncertainty has been the Achilles heel of the decision support community, since a practical system for problem solving should take the uncertainty of information or the uncertainty of the reasoning process into consideration. Naturally, we would like to have the mechanism in place for dealing effectively with both forms of uncertainty, even though all of the features may not be needed for a particular problem situation.

It is indeed one thing to map an entrée of beef into a wine descriptor in a deterministic sense, such as:

beef → *strong red meat* → *dry heavy red wine*

and quite another to introduce the notion of uncertainty, as in:

small lump -> {{{healthy, tests},0.6},{{healthy, tests, sick}, 0.4}}

The notation in the latter example is intended to demonstrate that a small lump is associated with the conditions of *healthy* or *tests* – are needed - with a certainty of 0.6 and or into the conditions of *healthy* or *tests* or *sick* with a certainty factor of 0.4, and this is before the inference process is introduced in the analysis. (Total comprehension is not required here, but it is clearly obvious that the presence of certainly complicates the situation.)

The stage is obviously set. An ADSS based on the advisor model should include the facility for handling uncertainty in order to truly represent the needs of modern decision makers.

Indicators and Evidence

In the wine and investment examples, there is relatively little risk in the advisory process, and this gives us the opportunity of focusing on the initial data base operation. The process represents a translation between the external world of the end user and the internal world of ADSS. In the wine advisor, for example, the indicator of beef is mapped into a set of attributes with which the system can work, as in:

beef -> strong red meat

or in generic form:

indicator -> meaning

The notion of an indicator is well-established in organized society in the form of symbols, variables, synonyms, and abbreviations. It is commonplace to encounter and indicator being used in everyday discourse, such as *successful,* which is commonly associated with a substantial set of characteristics, such as *aggressive, confident, intelligent,* and *focused.* Indicators can have different meanings to different people, however, which preserves the role of the domain specialist.

In the absence of uncertainty, and indicator has a definite meaning. In the presence of uncertainty, and indicator might mean something, and the degree to which a expert believes the assertion is known as its certainty value, or the certainty factor, as it is sometimes called. Thus, the indicator in the presence of uncertainty is a clue to the validity of an assertion; it becomes *evidence* when combined with its meaning and a certainty value or in its generic form:

Evidence: *indicator -> meaning* with *certainty value*

An obvious consideration applies to the notion of evidence. As in common everyday usage, a fact "is entered as evidence" only when it is relevant to the alternatives being considered in a decision situation.

END OF BLOCK # 1

DEALING WITH UNCERTAINTY – Block #2

One of the tenets of modern society is that things seem to get more complicated as time goes on. So when an unexpected event occurs, it is commonplace to hear something like, "That's the way life is nowadays." Most of our lives include built-in margins for various kinds of happenings that pop up unexpectedly. In reality, the margins can be reduced considerably by applying established technical methods to situations of uncertainty. Clearly, this type of analysis is probably not worth the effort in our personal lives. For enterprise activities, on the other hand, such an approach would seem to be the order of the day.

Certain phenomena are routinely associated with the occurrence of a particular event. For example, shortness of breath may be associated with a form of heart disease. Thus, given that form of heart disease, there is some probability that a patient with that disease will experience the symptom of shortness of breath. Sound like double talk, doesn't it. Similarly, the chance of a mechanical failure of a particular type – say a brake failure, fo example – is related brand of automobile. These are not unusual considerations, and they are normally sorted out during the course of everyday affairs. On the other hand, is the primary method of doing business for some forms of analysis, such as in diagnostic systems. This is the unique Bayesian viewpoint, however, that is most important and is precisely what permits the technical methods to be applicable to an information-rich environment.

The Bayesian form of analysis is so commonplace and so intuitively obvious that it hardly needs anything more than a brief mention. Running the risk of being excessively elementary and overly repetitive, therefore, here is the classic symptom-disease model. Given a specific disease, the occurrence of a specific symptom can be analyzed yielding the chance function that a patient with that disease will exhibit the symptom under consideration. Then by using a powerful method of inversion known as Bayes rule, the chance of contracting the disease given the symptoms can be calculated. The is the essence of the Bayesian form of analysis.

An information-based modality and the Bayesian viewpoint are a powerful combination in an advanced decision support system (ADSS). In short, indicators are associated with sets of possibilities, each of which is assigned a certainty factor The source of the uncertainty may be one of more of the following: inaccurate, incomplete, or incorrect information; different expert opinions;

concepts that are not well defined to start; and differing viewpoints of the decision situation.

END OF BLOCK #2

PRACTICAL CONSIDERATONS – Block # 3

In today's world, there is more than enough information to go around. The major challenge is to use it effectively and efficiently. The key point is the notion of an *indicator* that is a variable that is instantiated by assigning it a value. The proposition "the gross national product is down" and "the size of the lump is small" are indicators.

An indicator becomes *evidence* when an expert denotes its meaning to a particular problem domain and assigns a degree of belief to it. The evidence associated with the indicator "the gross national product is down," for example, could be that inflation will be moderate or high with a 0.6 measure of belief.

It is fortunately the case that several indicators can give some insight into the same basic problem domain. The is where the Dempster-Shafer Theory of Evidence is useful (see Information Block #2}. If the knowledge sources are independent, then Dempster's rule of combination can be used to fuse the evidence yielding a composite picture of the underlying state of nature. If one knowledge source (gnp=down) provides evidence of the form:

Inflation: {hi, mod}(0.6), {hi, low, mod, unc}(0.4)

And another (balance-of-payments=hi) provided:

Inflation: {mod, low}(0.7), {hi, low, mod, unc}(0.3)

Then an application of Dempster's rule would yield:

Inflation: {mod}(0.42), {hi, mod}(0.18), {low, mod}(0.28), {hi, low, mod, unc}(0.12)

In this example, the result is somewhat muddled, because the evidence is not confirmatory; clearly, the is the true nature of uncertainty. In many cases, the evidence is in fact confirmatory and insight into dealing with the result is straightforward. Two important and useful by-products of the methodology occur when the evidence is conflicting. The source of the differences can be readily identified and the analysis yields a consolidated position on the subject. These conclusions are obvious from the last example. In either case, the primary consideration involves what to do with the combined evidence, and this approach preserves the role of the specialist on the subject.

In a real sense, the key question is then "What do you do with the combined evidence?" The answer is quite simple. As in medical diagnosis, the end product of the analysis should be a prescription of what to do.

Using a knowledge-based modality of inquiry, the problem domain is mapped into a solution domain using a set of antecedent-consequent rules. In fact, there could easily be more than one solution domain for a given problem, and it is even possible for more than one problem domain to project into a single solution domain.

In the previous example, two obvious domains could be "portfolio" and "liquidity," expressed somewhat as follows:

Portfolio: {stocks}(0.252), {bonds, gold, stocks}(0.748)

Liquidity: {hi}(0.303), {low}(0.439), {hi, low}(0.257)

The reason that you might want to employ a solution domain is that it can be used to direct the activity of work object, taken in the is context to be persons or programs. This form of activity can be achieved through action items.

An *action item* is simply a *what to do,* which is based on an entry within a solution domain. Using a threshold of 0.4, for example, a decision script could be used to display the corresponding information, as in "Keep a low liquidity. Belief = 0.439." Other possibilities would be to initiate the execution of a program, activate an annunciator, and so forth.

REFERENCES

Gordon, J. and E. Shortliffe, "A Method of Management Evidential Reasoning in a Hierarchical Hypothesis," *Artificial Intelligence 26,* 323-357.

Kanak, L. and j. Lemer, *Uncertainty in Artificial Intelligence,* Amsterdam, 1986.

Katzan, H., "A Prototype Shell for the Modeling of Uncertainty," University of Fribourg # 133, 1987.

Katzan, H., *Managing Uncertainty,* New York, 1992.

Shafer, G., *A Mathematical Theory of Evidence, Princeton,* 1976.

END OF BLOCK # 3

END OF CHAPTER TWO

3

DEMPSTER-SHAFER THEORY OF EVIDENCE

Introduction

The Dempster-Shafer theory of evidence has commanded a considerable amount of attention in the scientific and business communities, because it allows a knowledge source to assign a numerical measure of belief to a proposition from a problem space, and provides a means for the beliefs accorded to independent knowledge sources to be combined. Dempster-Shafer theory is attractive because conflicting, as well as confirmatory, evidence from multiple agents may be combined.

The basis of Dempster-Shafer theory is a set of distinct and exhaustive possibilities called a *frame of discernment*. By convention, the frame of discernment is denoted by Q, such as in the following example:

Θ = {healthy, tests, sick}

Another way of viewing the concept is as follows: if q is a variable, then Θ represents the set of all possible values of q. The elements of a frame of discernment (usually called simply a "frame") corresponds to a proposition, such as:

"The true value of q is healthy"

Clearly, a knowledge source may assign a numerical measure of belief to a distinct element of Θ, which is equivalent to assigning a measure of belief to the corresponding proposition.

A measure of belief may also be assigned to a subset of Θ or to Θ itself. If A is a subset of Θ, such as:

$A = \{\text{test, sick}\}$

then assigning a belief function to A is equivalent to assigning that belief to the disjunctive proposition:

"The true value of A is *tests* or *sick*"

The capability of dealing with uncertain evidence that is restricted to a subset of the frame and at the same time being able to associate a measure of uncertainty to it gives Dempster-Shafer theory much of its intuitive and practical appeal.

Other than the ability to combine evidence, another important aspect of Dempster-Shafer theory is the manner in which complementary evidence is handled. One can assign a belief to a subset of Θ and leave the remainder of his or her belief uncommitted. For example, given the frame

$\Theta = \{\text{a, b, c, d}\}$

subset

$\text{A} = \{\text{a, c}\}$

and belief

$\text{Bel}(A) = \text{Bel}(\{\text{a, c}\}) = s$

then,

$$\text{Bel}(\Theta) = \text{Bel}(\{a, b, c, d\}) = 1\text{-}s$$

This is the same as saying, "I commit a belief of *s* to *A*, and I am unsure of the remainder of my belief, which I am leaving uncommitted." When a large number of alternatives are involved, as would normally be the case in everyday activities, this freedom is both powerful and convenient.

Frame of Discernment

The frame of discernment is the set of possibilities under consideration. Choosing an appropriate formulation for a problem is crucial, because the frame does not exist independently of the decision maker's knowledge. Representation is the creative aspect of using an advanced-support system (ADSS). In Dempster-Shafer theory, the elements of the frame acquire their meaning from what the analysts and decision makers know or believe they know.

By convention, the frame of discernment is demoted by Θ, as covered above. The set of all of the subsets of Θ is demoted by 2^{Θ}. For example, given the frame:

$$\Theta = \{a, b, c\}$$

the set of subsets of 2^{Θ} (i.e., the power set of Θ) is delineated as follows:

$$2^{\Theta} = \{\{a, b, c\},\ \{a, b\}, \{a, c\}, \{b, c\},\ \{a\}, \{b\},\{c\}\}$$

Because the null set yields not information, it is commonly omitted when delineating the contents of 2^{Θ}.

Support Functions

In Dempster-Shafer theory, and analyst apportions a unit it belief to an element of Θ. This belief represents a judgment as to the strength of the evidence supporting that proposition. When viewed in this manner, evidence focuses on the set corresponding to a proposition; this set is called a *focal set.*

The support for a focal set is a function *m* that maps an element of Θ denoted by *A*, onto the interval [0, 1]. The support function *m* is called a *basic probability assignment,* which is "assigned" to a proposition by an analyst or decision maker.

A support function is called a *simple support function* if it has at most one focal set not equal to Θ. A simple support function assigns a measure to belief to the focal set *A*, as follows:

$m(A) > 0$
$\mathrm{m}(\Theta) = 1\text{-}m(A)$, and
$m(B) = 0$, for all other subsets B of Θ

The simple support function for a focal set *A* assigns a portion of the total belief exactly to *A*, and not to its subsets or supersets. The remainder of the belief is assigned to Θ. Because the certainty functions must add up to 1, $m(\Theta)=1\text{-}m(A)$.

Consider, for example, the focal set

$G = \{\text{mod, low}\}$

corresponding to the proposition, "The true value of *G* is mod or low," over the frame of discernment

$\Theta = \{\text{hi, mod, low, unc}\}$

and the support function k. If k assigns a measure of belief to G as follows:

$$k(G) = 0.35$$
$$k(\Theta) = 1\text{-}k(G) = 0.65$$

then k is a *simple support function.*

It is possible that a body of evidence supports more than one proposition, as in the case that follows. If, for example

$$\Theta = \{a, b, c, d\}$$
$$A = \{a, b\}$$

and

$$B = \{a, c, d\}$$

Then the evidence supports two focal sets, which are A and B. If $m(A) = 0.5$ and $m(B)= 0.3$, then $m(\Theta) = 0.2$.

A support function with more than one focal set is called a *separable support function.* Separable support functions ae normally generated when simple support functions are combined.

The idea of combining simple support functions is a practical approach to the assessment of evidence. An analyst obtains information from a knowledge source, and it leads to an immediate conclusion – not with certainty, but with a certain level of belief. This is a normal straightforward means of handling human affairs and is precisely what people do. Then when additional information comes in, the various pieces of evidence are combined to obtain a composite picture of the situation.

Combination of Evidence

A method for combining evidence is known as Dempster's rule of combination. Evidence would normally be combined when it is obtained from two different observations, each with the same frame of discernment. The combination rule computes a new support function reflecting the scope of the combined evidence.

If m and n denote two support functions, then their combination is denoted by $m \oplus n$ and is called their *orthogonal sum*. (An orthogonal sum is computer by taking the intersection of the sets and the product of the certainty values.) As an example, consider he frame of discernment

$$\Theta = \{\text{healthy, test, sick}\}$$

and focal sets A and B based on two independent observations:

$A = \{\text{test, sick}\}$	$B = \{\text{healthy, tests}\}$
$m(A) = 0.8$	$n(B) = 0.6$
$m(\Theta) = 0.2$	$n(\Theta) = 0.4$

The entries are then combined as follows:

$$A \cap B = \{\text{tests, sick}\} \cap \{\text{healthy, tests}\} = \{\text{tests}\}$$
$$m(A) \times n(B) = 0.8 \times 0.6 = 0.48$$

$$A \cap n(\Theta) = \{\text{tests, sick}\} \cap \{\text{healthy, tests, sick}\} = \{\text{tests, sick}\}$$
$$m(A) \times n(\Theta) = 0.8 \times 0.4 = 0.32$$

$$B \cap m(\Theta) = \{\text{healthy, tests}\} \cap \{\text{healthy, tests, sick}\} = \{\text{healthy, tests}\}$$
$$m(B) \times n(\Theta) = 0.6 \times 0.2 = 0.12$$

$$m(\Theta) \cap n(\Theta) = \{\text{healthy, tests, sick}\} \cap \{\text{healthy, tests, sick}\}$$
$$= \{\text{healthy, tests, sick}\}$$
$$m(\Theta) \times n(\Theta) = 0.08$$

The entries are then combined as follows:

$$m \oplus \text{n}(\{\text{tests}\}) = 0.48$$

$m \oplus n(\{\text{healthy, tests}\}) = 0.12$
$m \oplus n(\{\text{tests, sick}\}) = 0.32$
$m \oplus n(\{\text{healthy, tests, sick}\}) = 0.08$
$m \oplus n = 0$, for all other subsets of Θ.

As another example, consider the same frame if discernment and the following focal sets:

$C = \{\text{healthy, tests}\}$	$D = \{\text{sick}\}$
$m(C) = 0.8$	$n(D) = 0.5$
$m(\Theta) = 0.2$	$n(\Theta) = 0.5$

The combination rule yields the following:

$m \oplus n(\{\varnothing\}) = 0.40$
$m \oplus n(\{\text{sick}\}) = 0.10$
$m \oplus n(\{\text{healthy, tests}\}) = 0.40$
$m \oplus n(\{\text{healthy, tests, sick}\}) = 0.10$

where the symbol $\varnothing$ denotes the empty set. The entry $m \oplus n(\{\varnothing\}) =$ 0.40 conflicting evidence The other values should then be normalized by dividing by (1-0.40) to obtain:

$m \oplus n(\{\text{sick}\}) = 0.167$
$m \oplus n(\{\text{healthy, tests}\}) = 0.667$
$m \oplus n(\{\text{healthy, tests, sick}\}) = 0.167$

where rounding accounts for the slight discrepancy.

Assignment of Belief

The assignment of belief involves the association of a number (i.e., the basic probability assignment) with the uncertainty inherent in a mutually exclusive and exhaustive set of possibilities. Sometimes, this assessment can be made directly from prior knowledge. In other cases, the determination of the measure of belief can be made by working sideways with the probability values in an associated problem.

Consider, for example, the task of determining the price of your favorite equity, the DS Corporation. Let *R* stand for the proposition "The value of DS stock will rise," and let *S* represent the proposition "Sam predicts the value of DS stock will rise." In this case, it would initially be desirable to have the conditional probability of *R* given *S*, but clearly, this sort of statistical information is not readily available from repeated trials, for obvious reasons. On the other hand, Sam is a crackerjack analyst, and given that he has performed a thorough fundamental analysis of a corporation, the stock will unquestionably rise of Sam says it will. The problem is that Sam is a busy guy, and this coupled with fact that Sam's ego sometimes gets in the way prompts Sam on occasion to give an investment opinion off the cuff, when he hasn't done his homework. Assume that Sam does his homework 70% of the time and let *H* represent the proposition "Sam has done his homework." Thus, $P(H)=0.70$ in this instance.

Recall here that we are looking for the probability of *R* and intend to use the values in an associated problem to infer it. Now if *S* is true, i.e., Sam predicts the value of DS stock will rise and also that Sam has done his homework, then events *H* and *R* will be compatible and we will have the following relationship:

$$M(\{R\}) = P(H) = 0.7$$

The other 30% of the tie, Sam makes an educated guess so anything is possible. Thus, the complement of *H* is compatible with the frame of discernment for the problem, so we have

$$m(\Theta) = 1\text{-}P(H) = 1\text{-}m(\{R\}) = 0.3$$

Clearly, the frame of discernment would ordinarily contain other possibilities, such as the value of DS stock will fall or be unchanged.

We can use a similar argument to obtain a secondary estimate of belief to be assigned to the frame of discernment. Suppose that by some quirk of fact, R is true when the value of the Dow rises. We can represent the latter proposition by D. Further, let us assume the Dow goes up 80% of the time. When the Dow doesn't rise, the fate of DS stock is another random walk down Wall Street. Thus we have the relationships

$$n(\{R\}) = P(D) = 0.8$$
$$n(\Theta) = 1\text{-}P(D) = 1\text{-}n(\{R\}) = 0.2$$

As before, D is only compatible with R and not with any other element of the frame of discernment.

Using the methods given earlier in this chapter, we can fuse this evidence using Dempster's Rule of Combination to obtain

$$m \oplus n(\{R\}) = 0.94$$

and

$$m \oplus n(\{\Theta\}) = 0.06$$

The process can be continued, so that if new evidence comes in, it can be readily combined with the latter result. In this case, recall that Θ represents the frame of discernment,

A Final Remark

Dempster-Shafer Theory can appear to be complicated. Once a reader understands, the combination of evidence is both convenient and powerful.

References

Gordon, J. and E. Shortliffe, "A Method of Management Evidential Reasoning in a Hierarchical Hypothesis," *Artificial Intelligence 26,* 323-357.

Kanak, L. and j. Lemer, *Uncertainty in Artificial Intelligence,* Amsterdam, 1986.

Katzan, H., "A Prototype Shell for the Modeling of Uncertainty," University of Fribourg # 133, 1987.

Katzan, H., *Managing Uncertainty,* New York, 1992.

Shafer, G., *A Mathematical Theory of Evidence, Princeton,* 1976

END OF CHAPTER 3

4

THE NEW CONNECTIONISM

INTRODUCTION

Artificial Intelligence is commonly regarded as the science of making machines do things that would require intelligence if performed by humans. This is not an intellectually useful definition, because it does not take the meaning of intelligence into consideration. For example, the task computing the payroll would require human intelligence if done by hand but is regarded as a commonplace data processing application. Of course, payroll is not a trivial application and neither are inventory management, online banking, and numerical computing – to name only a few cases. Well then, what characterizes and AI application? Moreover, is there a well-dividing line between AI and non-AI? The questions are largely rhetorical since it doesn't really matter how a particular computer application is classed. Artificial Intelligence is a pragmatic discipline, characterized more as a way of doing things than a specific well-defined technical concept. At least, that is what some people think. This chapter on Artificial Intelligence is intended to supply a background for the study of Dempster-Shafer theory, since the two disciplines are interrelated to some degree. Many AI problems are essentially solved through the application of Artificial Intelligence.

The AI Approach

It is possible to approach artificial intelligence from two points of view. Both approaches make use of programs and programming techniques. The first approach is to investigate the general principles of intelligence. The second is to study human thought, in particular.

Those persons engaged in the investigation of the principles of intelligence are normally charged with the development of systems that *appear* to be intelligent. This activity is commonly regarded as "artificial intelligence," which incorporates both engineering and science components.

Those persons engaged in the study of human thought attempt to emulate human mental processes to a lesser or greater degree. This activity can be regarded as "computer simulation," such that the elements of a relevant psychological theory are represented in a computer program. The objective off this approach is to generate psychological theories of human thought. The discipline is generally known as "cognitive science."

In reality, the difference between artificial intelligence and cognitive science tend to vary between "not so much" and "quite a lot" – depending upon the complexity of the underlying task. Most applications, as a matter of fact, contain elements from both approaches.

The Scope of AI

It is possible to zoom in on the scope of AI by focusing on the processes involved. At one extreme, the concentration is on the practicalities of doing AI programming, with an emphasis on symbolic programming languages and AI machines. In this context, AI can be regarded as a new way of doing programming. It necessarily follows that hardware/software system with AI components have the potential for enhanced end-user effectiveness.

At the other extreme, AI could be regarded as the study of intelligent computation. This is a more grandiose and encompassing focus with the objective of building a systematic theory of intellectual processes – regardless if they model human thought or not.

It would appear, therefore, that AI is more concerned with intelligence in general and less involved with human thought in particular. Thus, it may be contended that humans and computers are simply to species in the genus of information processing systems.

The Modern Era of Artificial Intelligence

The modern era of artificial intelligence effectively began with the summer conference at Dartmouth College in Hanover, New Hampshire in the summer of 1956. The key actors were Shannon from Bell Labs, Minsky from Harvard (later M.I.T.) McCarthy (later M.I.Y. and Stanford), and Simon from Carnegie Tech (later renamed Carnegie Mellon). The key results from the conference were twofold:

1. It legitimized the notion of AI and brought together a raft of piecemeal research activities.
2. The term "artificial intelligence" was coined and the name more than anything had a profound influence on the future direction of artificial intelligence.

The stars of the conference were Simon and Newell, who demonstrated the Logic Theorist – the first well-known reasoning program. They preferred the name, "Complex Information Processing," for the new fledgling science of the artificial. In the end, Shannon and McCarthy won out with zippy and provocative name, "Artificial Intelligence." In all probability, the resulting controversy surrounding the mane artificial intelligence served to sustain a critical mass of academic interest in the subject even during sporadic activity and questionable results.

One of the disadvantages of the pioneering AI conference was the simple fact that an elite group of scientists was created that would effectively decide "what AI is and what AI isn't," and how to best achieve it. The end result was that AI became closely aligned with psychology and not so much with neurophysiology, and to a lesser degree electrical engineering. AI became a software science with the main objective of producing intelligent artifacts. In short, it became a closed group, and this effectively constrained the field to a large degree.

In recent years, the direction of AI research has been altered somewhat by an apparent relationship with brain research and cognitive technology, which is known as the design of joint human-machine cognitive systems. Two obvious fallouts of the new direction are the well-known "Connection Machine" and the computer vision projects of the National Bureau of Standards in the United States.

Early Work on the Concept of artificial Intelligence

The history of AI essentially goes back to the Philosophy of Plato, who wrote that "All knowledge must be stateable in explicit definitions which anyone could apply," thereby eliminating appeals to judgment and intuition. Plato's student Aristotle continued in this noble tradition in the development of the categorical syllogism, which plays an important part in modern logic.

The mathematician Leibnitz attempted to quantify all knowledge and reasoning through an exact algebraic system by which all objects are assigned a unique characteristic number, whereby rules for the combination of problems would be established and controversies could be resolved by calculation.

The underlying philosophical idea was conceptually simple: Reduce the whole of human knowledge to a single formal system. The notion of formal representation has become the basis of AI and cognitive science theories since it involves the reduction of the totality of human experience to a set of basic elements that can be glued together in various ways.

To sum up, the philosophical phenomenologists argue that it is impossible to subject pure phenomena – i.e., mental acts which give meaning to the world – to formal analysis. Of course, AI people do not agree. They contend that "there is no ghost in the machine," and this is meant to imply that intelligence is a set of well-defined physical processes.

The discussion is reminiscent of the mind/brain controversy and it appears the AI perspective is that "the mind is what the brain does." Of course, the phenomenologists would reply that the definition of mind exists beyond the physical neurons; it also incorporates the intangible concepts of what the neurons do.

Accordingly, *Strong AI* is defined in the literature as the case wherein an appropriately programmed computer actually is a "mind." *Weak AI,* on the other hand, is the emulation of human intelligence, as we know it.

Intelligence and Intelligent Systems

There seems to be some value to the ongoing debate over the intelligence of AI artifacts. The term "artificial" in artificial intelligence helps us out. One could therefor contend that intelligence is natural if it is biological and artificial otherwise. This conclusion skirts the controversy and frees intellectual energy for more purposeful activity.

The abstract notion of intelligence, therefore, is conceptualized, and natural and artificial intelligence serve as specific instances. The subjects of understanding and learning could be treated in a similar manner. The productive tasks of identifying the salient aspects of intelligence, understanding, and learning emerge as a combined goal of AI and cognitive science. For example, the concepts of representation and reasoning, to name only two of many, have been studied productively from both artificial and biological points of view. Software products that are currently available can be evaluate on the basis of how well they can support the basic AI technologies.

The key question then becomes; How well do natural and artificial systems, as discussed above, match up to the abstract notion of intelligence?

Philosophical References

Boden, M., *Artificial Intelligence and Natural Man,* New York: Basic Books, Inc., 1977.

Edelson, T., "Can a system be intelligent if it never gives a damn?" *AAAI'86 Proceedings,* pp. 298-302.

Hofstadter, D., *Godel, Escher, Bach,* New York: Basic Books, Inc., 1979.

Quine, W., *Methods of Logic,* Cambridge: Harvard University Press, 1982.

Rappaport, W., "Philosophy, Artificial Intelligence, and the Chinese Room Argument," *Abacus (Summer 1986),* pp. 7-17.

Rose, W., *Into the Heart of the Mind: An American Quest for Artificial Intelligence,* New York: Harper and Row, Publishers, 1984.

Sowa, J., *Conceptual Structures,* Reading, Massachusetts: Addison-Wesley Publishing Company, 1984.

Woods, D., "Cognitive Technologies: The Design of Joint Human-Machine Cognitive Systems," *The AI Magazine,* Winter, 1986.

END OF CHAPTER 4

Part Two

APPLICATIONS OF UNCERTAINTY

5

IDENTITY ANALYTICS AND BELIEF STRUCTURES

INTRODUCTION

Identity is a major issue in the security of modern information systems and the privacy of data stored in those systems. Security and privacy concerns are commonly associated with behavioral tracking, personal-identifiable information (PII), the relevance of private data, data repurposing, identity theft, and homeland security. We are going to approach the subject from a data analytic viewpoint, where the primary challenge is to use identity in an effective way to determine group membership. Instead of focusing on the protection of identity in this paper, we are going to propose methods for using identity to make essential judgmental decisions.

Identity

Identity is a means of denoting an entity in a particular namespace and is the basis of security and privacy – regardless if the context is digital identification or non-digital identification. We are going to refer to an identity object as a *subject*. A subject may have several identities and belong to more than one namespace. A pure identity denotation is independent of a specific context, and a federated identity reflects a process that is shared between identity management systems. When one identity management system accepts the certification of another, a phenomenon known as "trust" is established. The execution of trust is often facilitated by a third party that is acknowledged by both parties and serves as the basis of digital identity in computer-based information systems and personal recognition is social affairs. (Salido 2010) There is another side to personal recognition. We are often afforded the identity of a person based on the judgment of a third party and are obligated to respond to that assessment. It would seem to be prudent in a civilized society to obtain additional information on the subject and combine the various items of information to obtain a composite view before engendering a timely response to the situation.

Privacy

Information systems typically process and store information about which privacy is of paramount concern. The main issue is identity, which serves as the basis of privacy or lack of it, and undermines the trust of individuals and organizations in other information-handling entities. The key consideration may turn out to be the integrity that organizations display when handling personal information and how accountable they are about their information practices. From an organizational perspective, control over information should remain with the end user or the data's creator with adequate controls over repurposing. From a personal perspective, a person should have the wherewithal to control his or her identity as well as the release of socially sensitive identity attributes. (Cavoukian 2009, 2010, ACLU 2010, CDD 2009, OECD 2010, FBI 2004) One of the beneficial aspects of the present concern over information privacy is that it places the person about whom data are recorded in proper perspective. Whereas such a person may be the object in an information system, he or she is regarded as the subject in privacy protection – as mentioned earlier. This usage of the word *subject* is intended to imply that a person should, in fact, have some control over the storage of personal information.

More specifically, the *subject* is the person, natural or legal, about whom data is stored. The *beneficial user* is the organization or individual for whom processing is performed, and the *agency* is the computing system in which the processing is performed and information is stored. In many cases, the beneficial user and the subject are members of the same organization.

The heart of the issue is privacy protection, which normally refers to the protection of rights of individuals. While the concept may also apply to groups of individuals, the individual aspect of the issue is that which raises questions of privacy and liberty. On the other hand, as in the case of terrorism and homeland security, privacy runs contrary to societal needs. We are going to keep those considerations in mind in this paper.

Belief

Belief is often regarded as a mental state in which a person holds a proposition to be true without necessarily being able to prove its truth to other persons. Even though absolute certainty is not required with belief, a person's set of beliefs can play an important role in the causation of behavior. Belief is associated with rational behavior and behavior that is not totally rational. Belief has a lot to do with a believer's mind. If a representation for belief P exists in a person's mind, then it is an *explicit belief*. If a representation for belief Q does not exist in a person's mind but is based on another proposition P, then it is an *implicit* belief. Beliefs that are based on an associative relationship are usually regarded as implicit beliefs.

Some authors class beliefs as being epistemic versus pragmatic and dispositional versus occurent. (Stanford 2010) With an *epistemic belief*, there is evidence for the belief. With *pragmatic belief*, there are practical reasons for the belief. Having been engaged in terrorist training, for example, would probably yield an epistemic belief that the subject has some inclination for terrorism. Pascal's argument to believe in God is an example of a pragmatic belief. It reads as follows: "The consequences of failing to believe in him if he exists (eternal fire and damnation) are much worse than the consequences of believing in Him if he does not exist (sin avoidance and contrition)."

Dispositional belief refers to the supposition that the subject is disposed to possess a certain stance on a topic or is inclined to a particular behavior. *Occurent belief* refers to the assumption that the subject is actually performing a sequence of actions. The penultimate example is also an example of dispositional belief. Direct knowledge, or information obtained from a trusted source, that a subject is performing a certain action is associated with occurent belief. In the latter case, verification of identity may be of some concern and be the difference between "belief in" and "knowledge of."

IDENTITY THEORY

The notion of identity is an important subject in philosophy, mathematics, and computer information systems. In its most general sense, identity refers to the set of characteristics that makes a subject definable. Each characteristic can be viewed as a single point in a three-dimensional Cartesian coordinate system where the axis are *subject, attribute,* and *value.* (Katzan 1975) Thus, the fact that George is twenty-five years old could be denoted by the triple <George, age, 25>. A set of characteristics over a given domain can uniquely identify a subject. This simple concept is the basis of identity and privacy in business, government operations, and everyday life. The notion of identity applies to organizational subjects as well as to personal subjects. An important aspect of modern identity theory is the linking of identity namespaces.

Knowledge and Power

The phrase "knowledge is power" is a popular means of expressing the value of information. So popular, in fact, that one would think its origin is the modern age of computers and information technology. That assumption, however, is not correct. The first reference that could be found is credited to the famous Sir Francis Bacon is his book, published in 1605, entitled *Advancement of Learning*, quoted as follows: (Bacon 1605)

> *But yet the commandment of knowledge is yet higher than the commandment over the will: for it is a commandment over the reason, belief, and understanding of man, which is the highest part of the mind, and giveth law to the will itself. For there is no power on earth which sets up a throne or chair of estate in the spirits and souls of men, and in their cogitations, imaginations, opinions, and beliefs, but knowledge and learning.*

Knowledge, in the sense that it is information concerning a thing or a person, can be used to further one's endeavors or it can be used to control a subject, thus diminishing its freedom and liberty. The protection of personal privacy is a Fourth Amendment right, and identity is the basis of privacy. The following sections give a philosophical view of identity.

Knowledge, Attributes, and Identity

Identity is primarily used to establish a relationship between an attribute or set of attributes and a person, object, event, concept, or theory. The relationship can be direct, based on physical evidence, and in other cases, the relationship is indirect and based on a reference to other entities. In a similar vein, the relationship can be certain or uncertain, and in the latter case, based on deduction or inference. The relationship determines an element of knowledge. For example, the knowledge element "you are in your car" is a statement in which "you" and "your car" are things that exist and the "in" is a relationship. Direct knowledge is known by *acquaintance* and is evidenced by a physical connection. Indirect knowledge is determined through a reference to a particular with which the analyst is acquainted. This form is known as knowledge by *description*. (Russell 1912) *Direct knowledge* is determined through sense data, memory, or introspection. *Indirect knowledge* is determined through a reference to another particular, as in "the person who ran for congress in 2004" or through a form of self-awareness where what goes on in subject's mind, for example, is estimated by an analyst's interpretation based on experience or self-evaluation.

Synthetic knowledge reflects certainty based on evidence inherent in the attribute values at hand. *Analytic knowledge* reflects a degree of uncertainty and is determined by deduction, as in "he is the only person with that 'attribute value'," or by inference based on known particulars, such as "all terrorists have beards." Inference, in this case, could be regarded as a form of derivative knowledge. The value of analytic knowledge is that it enables the analyst to exceed his or her limit of private experience. (Kant 1787) The concepts of knowledge, attributes, and identity are summarized in Table 1.

	Synthetic	***Analytic***
By acquaintance	A particular of which we have direct knowledge.	A particular of which we have knowledge based on deduction.
By description	A particular of which we have indirect knowledge by reference to a particular with which we are acquainted.	A particular of which we have indirect knowledge through inference (derivative knowledge).

Table 1. Elements of knowledge and identity.

Numerical and Qualitative Identity

Identity refers to the characteristics that make a subject the same or different. We are going to establish two forms of identity: numerical and qualitative. Two subjects are *numerically identical* if they are the same entity, such that there is only one instance. Two subject (or objects in this case) are *qualitatively identical* if they are copies or duplicates. In the popular movie *The Bourne Identity*, for example, the characters *Jason Bourne* and *David Web* are numerically identical, and the number of subjects is one. So it is with *Superman* and *Clark Kent* in another domain. On the other hand, a set of animals with the same biological characteristics – e.g., a species – is regarded as being qualitatively identical. The notion of qualitative identity is remarkably similar to the modern definition of a *category* informally defined as a collection of entities with the same characteristics, having the same values for the same attributes.

Theory of the Indiscernibles

An important aspect of identity theory is that subjects exhibit features of permanence and change, analogous to sameness and difference mentioned previously. We are going to discuss the concept of temporal identity in the next section. The notion of change implies a subject that undergoes a transformation and also possesses a property that remains unchanged. Both Locke and Hume have proclaimed that change reflects the idea of unity and not of identity. Leibnitz proposed the *Theory of Indiscernibles* suggesting that subjects (i.e., objects or entities) that are indiscernible are identical. (Stroll 1967) The subject of indiscernibles has implications for cloud computing, information systems, and change. To what extent a change in a characteristic denotes a change in identity is an open item at this time and implies that there is a probabilistic aspect to identity.

Russell approaches the subject of identity from an alternate viewpoint, analogous to definite and indefinite articles. Russell proposes that a description may be of two sorts: definite and indefinite.

A definite description is a name, and an indefinite description is a collection of objects x that have the property ø, such that the proposition øx is true. (Russell 1919) In the phrase *Dan Brown is a famous author*, for example, 'Dan Brown" is a name and the indefinite description is obvious, leading to the probabilistic link between a subject and a characteristic.

Temporal Identity

There is a rich quantity of philosophical literature on the change of identity over time. Are you the same person you were yesterday? Are there persistent attributes that allow for positive identity between time periods? As mentioned previously, entities in everyday life exhibit features of permanence and change. In the domain of personal identity, address attribute is a primary candidate for change. For example, John Smith lives at 123 Main Street. He moves out and another John Smith moves in. This is distinct possibility in a crowded city.

There is a form of *attribute duality* between a person subject and an object subject. A subject – an object, such as a residence, in this case – is characterized by who lives there. For example, rich people live on Sutton Place in New York. The discussion leads to four related concepts: endurant identity, perdurant identity, endurant attribute, and perdurant attribute. Clearly, the term *endurant* refers to a noun that does not change, where perdurant refers to one that does. Thus, the identity problem is essentially translated to an operant problem of "recognizing identity."

BELIEF STRUCTURES

We are going to assign subjects to an identity set based on values of attributes that characterize that set. An *identity set* is analogous to a namespace except that we are going to view identity from an analytic basis rather than from a privacy and security perspective. Consider the following scenario:

> *We are trying to identify subjects that belong to a certain group G. We know about the group G and its attributes. We have a paid knowledge source K_1 that informs us that subject A is a member of G. However, K_1 is not always correct, and we know that. We have used K_1 enough to know that he provides us with information when he needs money. We have an intuitive belief of how often he is correct. Fortunately, we have another source K_2 that can supply similar information. K_2 is not as hungry for money as K_1, and his opinion frequently runs contrary to K_1's. We would like to use analytics to combine the information from K_1 and K_2 so as to obtain a composite picture of the situation. Our resultant belief of A's membership in G is not the end of the story. The belief that we obtain of A's membership in G could then be propagated down the line to other analytic situations. However, we are going to go beyond the notion that even though subject A possesses G's attributes, it doesn't necessarily indicate that A is a member of identity set G.*

We are going to use belief structures, compatibility relations, consensus theory, and belief propagation to attack this problem. Consensus theory is a methodology for combining evidence based on Dempster-Shafer theory (Shafer 1976; Katzan 1992, 2006) and the mathematical combination of evidence (Dempster 1967). Consensus theory has commanded a considerable amount of attention in the scientific and business communities, because it allows a knowledge source to assign a numerical measure to a proposition from a problem space and provides a means for the measures accorded to independent knowledge sources to be combined. Consensus theory is attractive because conflicting, as well as confirmatory, evidence from multiple sources may be combined.

Frame of Discernment

A frame of discernment is a means of representing the possibilities under consideration, as in the following example:

V = {medicine, law, education}

Clearly, the elements in a frame of discernment are, in fact, propositions that can be interpreted as events or states. Thus, if component s_i of system *S* over domain *V* were associated with the symbol **law,** then that state is equivalent to the proposition, "The true value of *V* for component s_i is **law,**" or in ordinary language, "s_i prefers **law.**" Accordingly, the set S of propositions S_i, $S = \{S_1, S_2, \ldots, S_n\}$ represents the collection of states of a system under analysis. Clearly, at an agreed upon point in time, one proposition is true and the others are false.

The basis of identity analytics is a frame of discernment (Θ). Accordingly, a knowledge source may assign a numerical measure to a distinct element of Θ, which is equivalent to assigning a measure of belief to the corresponding proposition. In most cases, the numerical measure will be a belief assignment. A measure of belief may also be assigned to a subset of Θ or to Θ itself. Consider a frame of discernment Θ and its power set denoted by 2^{Θ}. Given the frame $\Theta = \{a, b, c\}$, its power set is delineated as: $2^{\Theta} = \{\{a, b, c\}, \{a, b\}, \{a, c\}, \{b, c\}, \{a\}, \{b\}, \{c\}, \{\varphi\}\}$. In identity analytics, a knowledge source apportions a unit of belief to an element of 2^{Θ}. This belief can be regarded as a mass committed to a proposition and represents a judgment as to the strength of the evidence supporting that proposition. When viewed in this manner, evidence focuses on the set corresponding to a proposition; this set is called a *focal set*. The support for a focal set is a function *m* that maps an element of 2^{Θ}, denoted by *A*, onto the interval [0,1]. Given a frame of discernment Θ and function m: $2^{\Theta} \rightarrow [0,1]$, a support function is defined as: $m(\varphi) = 0$, where φ is the null set, $0 \leq m(A) \leq 1$, and $\sum m(A) = 1$.

A simple support function assigns a measure of belief to the focal set A, as: $m(A)>0$; $m(\Theta)=1-m(A)$; and $m(B)=0$, for all $B\subset 2^{\Theta}$ and $B\neq A$. The simple support function for a focal set A assigns a portion of the total belief exactly to A and not to its subsets or supersets. The remainder of the belief is assigned to Θ, because certainty function must add up to 1, $m(\Theta) = 1-m(A)$. It is possible that a body of knowledge or evidence supports more than one proposition, as in the following case. If $\Theta = \{a, b, c, d\}$, $A = \{a, b\}$, and $B = \{a, c, d\}$, then the evidence supports two focal sets, which in the example, are A and B. If $m(A) = 0.5$ and $m(B) = 0.3$, then $m(\Theta) = 0.2$. A support function with more than one focal set is called a *separable support function*. Separable support functions are normally generated when simple support functions are combined. The notion of combining simple support functions is a practical approach to the assessment of evidence. An analyst obtains information from a knowledge source, and it leads to an immediate conclusion – not with certainty, but with a certain level of belief. This is a straightforward means of handling human affairs and is precisely what people do when analyzing situations in everyday life. If additional information comes in, the various pieces of evidence are combined to obtain a composite picture of the situation.

Compatibility Relations

In this particular instance, we are going to establish relations between three sets and the frames of discernment for K_1, K_2, and A, where the K_i are the knowledge sources and A is the subject. The relations will be represented as:

$K_1 \rightarrow A$
$K_2 \rightarrow A$

and the frames:

$A = \{m, n\}$
$K_1 = \{r, u\}$
$K_2 = \{c, i\}$

The question is whether A is a member of G, denoted by m, or not a member of G, denoted by n. As far as K_1 is concerned, he might be telling us what he thinks we want to hear, so his judgment is classed as reliable, denoted by r, or unreliable, denoted by u. K_2 is simply correct or incorrect, denoted by c or i, respectively.

We can now establish the requisite compatibility relations, based on the fact that K_1 informs us that A is a member of G, and K_2 informs us that A is not a member of G.

1. If K_1 has based his opinion on credible evidence and is operating in a trustworthy manner, then he is in state r that is compatible with state m for A. If K_1 just needs the money or doesn't have good evidence, then he is in state u that is compatible with both states m and n. Thus, we have the compatibility relation:

   $\{(r, m), (u, m), (u, n)\}$

2. If K_2 is behaving as normal, and there is no reason at this point not to accept that, then he is in state c that is compatible with state n for A. If K_2 is in state i then all bets are off, and this state is compatible with A's states m and n. We then have the second compatibility relation, which is:

   $\{(c, n), (i, m), (i, n)\}$

Compatibility relations will allow us to assign belief to the assertions of K_1 and K_2 and propagate that belief through the belief network, resulting in a set of focal sets that can be combined using Dempster's rule in order to obtain a composite picture of the situation. Up to this point, we are working in the problem space for the analysis.

Prior Belief

An analyst assigns a measure of credibility to a knowledge source. In our example, let the belief assigned to K_1 be denoted by p and the belief assigned to K_2 be denoted by q, yielding the following prior belief:

Source	*Belief*
K_1	$\{[(r), p]. [(r, u), 1\text{-}p]\}$
K_2	$\{[(c), q]. [(c, i), 1\text{-}q]\}$

Since we are in the problem space, our belief in K_1 and K_2 is invariant.

Belief Propagation

Belief propagation transfers the knowledge from the problem space to the solution space using the compatibility relations, resulting in the following focal sets:

Source	*Focal Set*
K_1	$\{[(m), p]. [(m, n), 1\text{-}p]\}$
K_2	$\{[(n), q]. [(m, n), 1\text{-}q]\}$

The results of belief propagation assign the mass of the information received from K_1 to (m) and the remainder of the belief is assigned to (m, n), which is the frame, denoted by Θ in the above introduction. A similar argument applies to K_2 such that the mass of that belief is assigned to (n) and Θ, respectively.

Combination of Evidence

Using Dempster's rules of combination (Dempster op cit.), the resulting focal sets can be combined yielding the following assessment in the solution space:

$$\left[(m),\frac{p(1-q)}{1-pq}\right],\left[(n),\frac{(1-p)q}{1-pq}\right].\left[(m,n),\frac{(1-p)(1-q)}{1-pq}\right]$$

using symbolic math from calculations in *Mathematica*™. Applying the expression to several values of p and q yields the following results:

K_1 (p)	K_2 (q)	$K_1 \oplus K_2$
.6	.7	{[(m), 0.310], [(n), 0.483], [(m, n), 0.207]}
.7	.8	{[(m), 0.318], [(n), 0.545], [(m, n), 0.136]}
.8	.9	{[(m), 0.286], [(n), 0.643], [(m, n), 0.071]}
.7	.5	{[(m), 0.538], [(n), 0.231], [(m, n), 0.231]}

This is what we wanted to show. QED.

Summary

We have introduced the theory of identity and applied it to the combination of knowledge for assessment of whether a subject is a member of a certain group. We have introduced belief structures and a relevant methodology for mapping a problem space into a solution space.

Appendix: Combination of Evidence

A method of combining evidence is known as Dempster's rule of combination (Dempster 1967). Evidence would normally be combined when it is obtained from two different observations, each over the same frame of discernment. The combination rule computes a new support function reflecting the consensus of the combined evidence.

If m_1 and m_2 denote two support functions, then their combination is denoted by $m_1 \oplus m_2$ and is called their *orthogonal sum*. The combination $m_1 \oplus m_2$ is computed from m_1 and m_2 by considering all products of the form $m_1(X) \bullet m_2(Y)$, where X and Y range over the elements of Θ; $m_1(X) \bullet m_2(Y)$ is the set intersection of X and Y combined with the product of the corresponding probabilities.

For example, consider the frame of discernment

$$\Theta = \{h, t, s\}$$

and views A and B, based on two different observations over the same frame:

$$X = \{\{h\},0.6\},\{\{t\},0.3\},\{\{s\},0.1\}\}$$
$$Y = \{\{h\},0.4\},\{\{t\},0.4\},\{\{s\},0.2\}\}$$

The entries are combined, as follows, using Dempster's rule of combination:

$$m_1 \oplus m_2(\{h\}) = 0.24$$
$$m_1 \oplus m_2(\{t\}) = 0.12$$
$$m_1 \oplus m_2(\{s\}) = 0.02$$
$$m_1 \oplus m_2(\{\emptyset\}) = 0.62$$

Thus, for $A_i \cap B_j = \mathbf{A}$ and $m_1 \oplus m_2 = \boldsymbol{m}$, the combination rule is defined mathematically as:

$$\boldsymbol{m}(\mathbf{A}) = \sum m_1(A_i) \bullet m_2(B_j) / (1 - \sum m_1(A_i) \bullet m_2(B_j))$$

The denominator reflects a normalization process to insure that the pooled values sum to 1. So, in this instance, the normalization process yields the combination

$$X \oplus Y = \{\{h\},0.63\},\{\{t\},0.32\},\{\{s\},0.05\}\}$$

after normalization by dividing the combined assessment by (1-0.62) or 0.38. Because the problem is well-structured, the representation can be simplified as

$$X \oplus Y = \{0.63,0.32,0.05\}$$

For views A = $\{A_1,A_2,\ldots,A_n\}$ and B=$\{B_1,B_2,\ldots,B_n\}$, the combination rule can be simplified as

$$A \oplus B = \{A_1 \times B_1/k, A_2 \times B_2/k, \ldots, A_n \times B_n/k\}$$

where

$$k = \sum_{i=1}^{n} A_i \times B_i$$

We will refer to latter the equation as the *simplification rule*. (Katzan 2009) Readers are directed to Shafer (1976) and Katzan (1992) for additional information on Dempster's rule of combination.

REFERENCES

1 Bacon, Sir Francis. 1605. *Advancement of Learning.* (Republished in the *Great Books of the Western World.* Volume 30, Robert Maynard Hutchins, Editor in Chief, Chicago: Encyclopedia Britannica, Inc., 1952).

2 Black, M. 1952. Identity of Indiscernibles. *Mind* 61:153. (Secondary reference.)

3 ACLU of Northern California. 2010. *Cloud Computing: Storm Warning for Privacy?* www.dotrights.org, (downloaded 3/11/2010).

4 Cavoukian, A. 2009. *Privacy in the Clouds.* Toronto: Information and Privacy Commission of Ontario (www.ipc.on.ca).

5 Cavoukian, A. 2010. 7 Laws of Identity: The Case for Privacy-Embedded Laws of Identity In the Digital Age." Toronto: Information and Privacy Commission of Ontario (www.ipc.on.ca).

6 Center for Digital Democracy (CDD). 2009. *Online Behavioral Tracking and Targeting: Legislative Primer September 2009.* www.democraticmedia.org/privacy-legislative-primer. (downloaded 3/11/2010).

7 Dempster, A.P. 1967, "Upper and Lower Probabilities Induced by a Multivalued Mapping," *The Annals of Statistics* 28:325-339.

8 Federal Bureau of Investigation. 2004. *Privacy Impact Assessment.* www.fbi.gov/biometrics.htm. (downloaded 2/20/2010).

9 Kant, I. 1787. *Critique of Pure Reason.* (Republished in *Basic Writings of Kant.* Allen W. Wood, Editor, New York: The Modern Library, 2001).

10 Katzan, H. 1975. *Computer Data Management and Data Base Technology,* New York: Van Nostrand Reinhold Co.

11 Katzan, H. 1992. *Managing Uncertainty: A Pragmatic Approach,* New York: Van Nostrand Reinhold Co.

12 Katzan, H. 2008. Categorical Analytics Based on Consensus Theory. *Journal of Business and Economics Research,* 6(8), 89-102.

13 Katzan, H. 2010. On the Privacy of Cloud Computing. *International Journal of Management and Information Systems,* (accepted for publication).

14 OECD 2010. OECD Guidelines on the Protection of Privacy and Transborder Flows of Personal Data. www.oecd.org. (downloaded 3/23/2010).

15 Russell, B. 1912. *The Problems of Philosophy.* (Republished by Barnes & Noble, New York, 2004).

16 Russell, B. 1919. *Introduction to Mathematical Philosophy.* (Republished by Barnes & Noble, New York, 2005).

17 Salido, J. and P. Voon. 2010. A Guide to Data Governance for Privacy, Confidentiality, and Compliance: Part 1. The Case for Data Governance. Microsoft Corporation.

18 Shafer, G. 1976, *A Mathematical Theory of Evidence,* Princeton, NJ: Princeton University Press.

19 Stroll, A. 1967. *Identity.* (Entry in *The Encyclopedia of Philosophy,* Volume 4, Paul Edwards, Editor-in-Chief, New York: Macmillan Publishing Co., 1967).

END OF CHAPTER 5

6

EVOLUTIONARY DYNAMICS OF SERVICE PROVISIONING

INTRODUCTION

It is well established that service is a relationship between a provider and a client that captures value. (Spohrer 2007, Katzan 2008, Lusch 2008) It has also been recognized that service is a process, and that all products may in fact be services. (Vargo 2009) This perspective gives us a wide-open playing field for investigating the evolutionary dynamics of service provisioning in particular areas of concern. The purpose of this paper is to develop a model (i.e., a set of equations and relationships) that describes how service evolves through a metaphor of selection, survival of the fittest, replication, and mutation. We will be dealing with categories of service providers, because as Martin Nowak aptly puts it, "Evolutionary dynamics acts on populations. Neither genes, nor cells, nor individuals evolve; only populations evolve." (Nowak 2005) In fact, the model has implications for service provisioning within a service collective. (Katzan 2010b) A client selects a service provider within a particular category for a variety of reasons that pertain to the overall problem domain. As conditions change, the client may select another provider in the same category or switch to a provider in another category. In fact, the client may leave the service system altogether. This paper seeks to apply evolutionary dynamics to a population of service elements. In this instance, the service system responds to a population of clients that serves as the environment in a metaphor of service evolution.

Evolutionary Service Elements

You don't have service if you don't have clients, also known as customers or consumers, depending upon the context. Clients often have a choice among providers over a particular service domain. Some everyday examples will set the scene. In the medical area, for example, a patient could choose from the categories of *medical doctor, osteopathic physician,* or *chiropractor.* This might be regarded as a pure service, as would be the area of education composed of private school, public school, and tutoring. An example of the domain of "services that involve a product" could be a fast-food restaurant offering hamburgers, fish sandwiches, chicken sandwiches, and salads. A bakery providing bread, donuts, muffins, and cake would represent examples of a "product as a service," as would be an automobile company. A familiar means of representing a set of categories is a frame of discernment, as in the following list:

Medical	=	{medical doctor, osteopathic physician, chiropractor}
Fast food	=	{hamburger, fish sandwich, chicken sandwich, salad}
Bakery	=	{bread, donut, muffin, bagel, cake}
Automobile	=	{sedan, coupe, convertible, van, suv}
Education	=	{public, private, tutor}

The key point is that the client has a free choice within in a service domain. Usually, the choice may change but not the domain. Thus, we have populations of service providers categorized by definitive attributes. We are going to study the evolution of the categories, as a biological scientist might study a species. Within a specific domain, members of a category have common attributes with enough variety to allow for evolutionary service elements, such as natural selection, replication, and mutation.

Elementary Operations

A client enters and leaves a *service system*, regarded as a collection of resources, economic entities, and supplementary services capable of engaging in and sustaining one or more service events. (Katzan 2010a) The client selects a service provider from among providers in a relevant service category that possesses sufficient requisite variety to enable that choice. (Ashby 1964) For example, a client may choose a particular physician from a medical-doctor category over a *medical* domain, a laptop computer over a *computer* domain, a kind of hamburger over a *fast food* domain, or a kind of vehicle over an *automobile* domain. We are concerned about the result of client behavior in such situations.

Within a broad service domain, such as fast food or medical treatment, the client has the choice of behaviors, such as the following:

Stay with that service category (perhaps comprised of diverse providers)
Change category
Leave the service system altogether

A service provider category encapsulates multiple services in the following manner: (1) A single provider performs multiple services; (2) A client can use a single provider for different service events; and (3) A client can move between providers in the category.

There is customarily movement by clients between service categories depending upon social and other metaphysical considerations, for which there is normally a cost and benefit accrued by changing categories. An organization, for example, may change consulting services or cloud service providers. Some categories within a service domain are more attractive for a variety of reasons, collectively identified as "fitness." When clients select a category of service, they are exercising the right to choose, known as *democratization*. Thus, the process of independent choice, otherwise known as selection, contributes to the fitness of a category with the underlying assumption that clients select the most attractive service provider, based on their temporal needs.

Service dynamics acts on populations of service providers within a particular service domain, so we will be viewing service through a service category lens. Service categories effectively "steal" from other categories by being more fit for selection and replication. The precise reasons that clients move between categories are not relevant to this form of analysis. The important thing is that they do it based on evolutionary considerations. In fact, even the client set is not important for this analysis, since, as mentioned, it serves as the environment for the provider set. We are solely interested in the kinematics of the behavior of populations of service providers.

Biological Metaphor

In this manner of analysis, there are two populations: a set of clients and a set of providers. A client selects from the available providers. We are not necessarily concerned with individual clients, except for the fact that they select a service category – of course, over a service domain with an associated frame of discernment.

We are going to refer to the collection of service providers in a service domain as a *population*, comprised of service categories, referred to a *species*. Species evolve and attract more clients. Species also regress and lose clients. Evolution progresses through selection and replication. Regression can be viewed as the result of a lack of fitness.

Fitness is a complex phenomenon linked to the ability of a service category's ability to attract clients and evolve in a purposeful manner. An important component of fitness is a service metric, such as market share, sales volume, or revenue. This subject is covered in the next section.

Service categories further evolve through combination and division. New service categories can be formed by combining two or more constituent categories, sometimes known as merging. Service division is a process by which an established service category is divided to form competing categories or supply supplementary services.

As in biological processes, the essence of service dynamics is competition. Categories compete to enhance their values of the service metric and to sustain their position in the chain of categories in a service collective. (Katzan 2010b) As a focal point, one may regard provider behavior as reflecting "market share" and "sustainability."

Democratization And Monetization Of Services

In the marketplace of services, providers compete for market share, volume, or revenue because there is a limited resource pool (scarcity) or a controlled growth in the availability of required resources. The selection process is a measure of fitness. The clients essentially vote for the provider that is most fit by selecting a specific service category, known as the *democratization of services.*

Once a service metric is selected, it is important to recognize a dependency on whether the population is fixed or variable. If the birth rate (***b***) and death rate (***d***) of clients is equal, i.e., ***b=d,*** the methodology, related to fitness, is adjusted from that of a dynamic population, where ***b>d*** or ***b<d***. For example, if the purchase ability of a set of vehicle buyers is fixed, e.g., one vehicle per family, then a move from sedan to utility vehicle, for example, changes the service metric from the case where the client supplements a sedan with a utility vehicle.

Monetization of services refers to the "survival of the fittest" notion, such that a provider category can adjust its service provisioning in an attempt to change its market position. The form of modeling presented in this paper necessarily applies to both the democratization and monetization of service.

EVOLUTIONARY BEHAVIOR

Following Nowak (2006) and Olfati-Saber (2007), we seek to develop a formal description of client behavior as elements move between service providers over a frame of discernment. The provider network can be viewed as a complete graph G=(N, E), where the nodes (N) represent the case where a set ;of clients enlist the services of a particular set of service providers, and the edges (E) represent movement between the providers resulting from socioeconomic conditions.

Service Demographics

The methods of analysis are based on a collection P of n service provider groups over an area of endeavor A. The expectation is that a client will select of member of P for service. For example, a patient will choose among the distinct categories of medical doctors, osteopathic physicians, or chiropractors for treatment. Each category may incorporate one of more providers. Similarly, a customer of a bakery shop may choose a donut, muffin, or bagel – using the "all products are essentially services" (Vargo 2009) conjecture – where in fact there are several choices on the shelf.

Represent the categories of population P over domain A as the vector

$$C = [c_i],\ i=1, 2, \dots, n$$

The proportion of the client population that selects a particular provider category is denoted by the vector

$$X = [x_i],\ i=1, 2, \dots, n$$

where x_i represents the proportion of the client population associated with provider category c_i, where $\sum_i x_I = 1$. The analyst may interpret the metric X in any manner pertaining to the analysis, such as the number of clients, total value of client interactions, or the volume of client service.

The attitudes and beliefs resulting from prior service event or interactions with other clients affect the x_i. At this stage of the research, we are not going to consider client movement within a category.

Applying the biology metaphor, the collection of provider categories C are a species in the population P, and a category replicates in response to client requirements – or more accurately, client preferences.

An example of this form of replication would be innovation and competition based on increasing market opportunities. The notion of fitness applies here as the most-fit providers attract the most clients.

Service Kinematics

The status of a service system, under consideration at a given point in time, is therefore represented by the state vector:

$$X = [x_1, x_2, \ldots, x_n]$$

where each x_i reflects the proportion of clients that choose provider category c_i. Consider a reward matrix $A = [a_{ij}]$ that represents the reward that a client would obtain by switching from category c_j to category c_i. Thus, a service category replicates in response to client preferences. One could conceptualize a business switching for computer consulting services from one provider to another, but only if there is some measurable utility for doing so. When a provider c_i replicates, the value of its state value x_i increases based on a fitness measure determined by the values in the reward matrix.

Mutation

Mutation is an important element in replication based on fitness. The probability at clients in group c_i mutates to group c_j is given by matrix:

$$Q = [q_{ij}], \text{ where } 0 \leq q_{ij} \leq 1$$

Q is square row stochastic matrix of order $n \times n$ and $\Sigma_j^n q_{ij} = 1$. Thus, q_{ij} can be regarded as the rate of transference from x_i to x_j based on exogenous conditions.

SERVICE DYNAMICS

Several equations, adapted from mathematical biology, Nowak (op cit.) and Olfati-Saber (op cit.), are used to describe client behavior. In order to properly frame the problem of service dynamics, several considerations are useful for explaining the terminology adapted from evolutionary biology. As with most instances of socioeconomic behavior, some service providers dominate because there is a positive utility for their selection. The term commonly used to reflect a position is "fitness." The service providers that are most fit receive the attention, and not-so-fortunate providers adapt to better compete and increase their sustainability index. We are going to base the fitness metric on the reward matrix A, covered earlier. Mutation is also introduced, because even though clients are assumed to be rational, other considerations from the environment often apply.

Fitness

Fitness is a measure of an element c_i from category C based on the reward matrix A. The fitness vector is $F = [f_i]$, where i=1, 2, ..., n, and is computed as:

$$f_i = \Sigma_j a_{ij} x_j \qquad [1]$$

and its average Φ for category C is computed as:

$$\Phi = \Sigma_i x_i f_i \qquad [2]$$

Clearly, the notion reflects the reward of switching from c_j to c_i.

The types c_i are competing for resources from the environment of clients, where for example, one service category evolves faster, based on the reward matrix for clients and out-competes other service providers.

The basic measure of fitness is the degree to which a category attracts clients, represented by the column *i* of the matrix, resulting in equation [1].

Service Evolution

When $f_i > \Phi$, the state value x_i for c_i increases and when $f_i < \Phi$ the state value x_i for c_i decreases. The associated evolution value for x_i is computed from the following difference equation:

$$new\ x_i = x_i\ (1+(f_i - \Phi)) \qquad [3]$$

where f_i is the fitness of c_i and x_i is the corresponding state value. The state value x_i is the proportion of clients that choose provider c_i. (Novak 2006, Olfati-Saber 2007) An example of service evolution is given in a succeeding section.

Example of Service Evolution

The example describes an a service domain comprised of computer vendors, given by the frame of discernment C:

C = {Apple, Sony, Dell}

and the state vector X, as:

$$X = (0.5, 0.2, 0.3)$$

such that c_1 = Apple corresponds to a state value of x_1=0.5. A hypothetical reward matrix is given as:

$$A = \begin{pmatrix} 1 & 1.5 & 2 \\ 0.5 & 1 & 0.25 \\ 0.25 & 0.5 & 1 \end{pmatrix}$$

Each row represents the rewards of migrating to a category (i.e., the row) from the other categories (i.e., the column). Thus, a client would obtain a reward of 1.5 from Sony to Apple and of 2 from Dell to Apple, and so forth. Row one represents Apple, row two represents Sony, and row represents Dell. The fitness vector $F = [f_i]$ is determined as

$$F = (1.4, 0.35, 0.525)$$

from equation [1], where Φ is computed as 0.82 from equation [2]. Accordingly, the new state vector, referred to as *new X*, is computed from equation [3], as

$$\text{new}\, X = (0.79, 0.106, 0.12)$$

reflecting an evolution to the succeeding state.

Mutation Matrix

A form of mutation, representing as conversion between states, is modeled as the rpw stochastic matrix $Q = [q_{ij}]$ such that $\Sigma_j q_{ij}=1$. The mutation matrix is computed from a weight matrix $W = [w_{ij}]$, where $w_{ij} = a_{ij}/(\Sigma_j a_{ij})$. A mutation parameter μ is defined, so that $q_{ij} = \mu w_{ij}$. (Olfai-Saber cp cit.)

Introducing mutation into the modeling, we can develop a more sophisticated replication model as

$$new\ x_i = x_i\ (1+((\Sigma_j x_j f_j q_{ji})- \Phi x_i))$$

Mutation is beyond the scope of this paper and presentation, but the notion of mutation is an intriguing topic for evolutionary modeling.

SUMMARY

The subject of evolutionary modeling based on the work of Nowak and Olfati-Saber has been introduced, and its application to the evolutionary modeling of service provisioning has been presented. Equations for fitness and replication have been developed that differ from biological processes, but retain the requisite elements for a viable methodology. A relevant numerical example has been included.

REFERENCES

Ashby, W. 1964. *An Introduction Cybernetics.* London: Chapman and Hall.

Katzan, H. 2008. *Service Science: Concepts, Technology, Management.* New York: iUniverse, Inc.

Katzan, H. 2010a. Service Analysis and Design. *International Applied Business Research Conference Proceedings,* Orlando, FL: IABR Conference, January 4-6, 2010.

Katzan, H. 2010b. Service Collectivism, Collaborations, and Duality. *International Applied Business Research Conference Proceedings,* Orlando, FL: IABR Conference, January 4-6, 2010.

Lusch, R., Vargo, S., and G. Wessels. 2008. Toward a conceptual foundation for service science: Contributions from service-dominant logic. *IBM Systems Journal,* 47(1): 5-14.

Nowak, M. 2006. *Evolutionary Dynamics: Exploring the Equations of Life.* Cambridge, MA: The Belknap Press of Harvard University Press.

Olfati-Saber, R. 2007. Evolutionary Dynamics of Behavior in Social Networks. *Proceedings of the 46th IEEE Conference on Decision and Control* (December 2007).

Spohrer, J., Vargo, S., Caswell, N. and P. Maglio. 2007. The Service System is the Basic Abstraction of Service Science. *IBM Almaden Research Center,* http://www.almaden.ibm.com/asr.

Vargo, S. and M. Akaka. 2009. Service-Dominant Logic as a Foundation fro Service Science: Clarifications. *Service Science,* 1(1): 32-41.

END OF CHAPTER 6

7

CATEGORICAL ANALYTICS BASED ON CONSENSUS THEORY

INTRODUCTION

A system can be viewed generically as a collection of elements that interoperate to achieve a given purpose. Each of the elements is in one of a set of permissible states. The persons in a legislature, the components in a mechanical or electrical assembly, and the elements in a biological organism are typical examples of systems. In voting systems, cliques are often formed that span groups and subgroups. A constituency and a variety of intrinsic factors, such as personal economics, education, gender, feelings about certain moral issues, and so forth, influence individuals in voting systems. One viewpoint is that a voter is subjected to collection of rational forces from which a consensus must be drawn. Lastly, the state of various exogenous factors, such as the economy and the world's political condition, could also be factored into an analysis that is focused on the above considerations. The concepts apply equally well to investors in the financial community, as well as to actors in a variety of design and analysis problems.

ANALYSIS

A *frame of discernment* (hereafter referred to as simply a *frame*) is a set of mutually exclusive and collectively exhaustive possibilities for the state of a component of a system within a particular problem domain (referred to as the *domain*). Three obvious examples of frames are:

V = {McCain, Obama, Nader}
M = {For, Undecided, Against}

In general, a frame is a means of representing the possibilities under consideration.

Clearly, the elements in a frame are in fact propositions that can be interpreted in the usual fashion as events or states. Thus, if component s_i of system *S* over domain *V* were associated with the symbol "McCain," then that state is equivalent to the proposition, "The true value of *V* for component s_i is *McCain,*" or in ordinary language, "s_i prefers or has voted for *McCain.*"

The set S of propositions S_i,

$$S = \{S_1, S_2, \ldots, S_n\}$$

commonly represents the collection of states of a component of a system under analysis. Clearly, at an agreed upon point in time, one proposition is true and the others are false. If proposition S_i is true, then we can assert that

S_i = T and
S_j = F, where $i \neq j$

where T and F represent "true" and "false," respectively.

Uncertainty

Prior to an agreed point in time (τ), we obviously do not know the state of the system under analysis or its components with any degree of certainty. The expectation that a part of the system will be in a particular state at time τ is denoted by a real number $p(S_i)$ associated with each of the propositions in the frame $S = \{S_i\}$, $i=1,2,\ldots,n$, such that

$$0 \leq p(S_i) \leq 1$$

and

$$\sum_{i=1}^{n} p(S_i) = 1$$

It follows that the function p can be extended to the lattice $\underline{S}$ of S such that

$$p(k) = \sum_{i=1}^{k} p(S_i) = 1$$

where k is an element of $\underline{S}$. This is simply the addition rule for mutually exclusive events.

Information Theory

If the likelihood of an event is relatively remote and it happens to turn out to be true at an appointed time, we are surprised. Similarly, if the chance of the same event is very likely and it comes out true, we are not overly surprised. The extent of our surprise represents our "information gain" in the sense that if we knew that information, we wouldn't be surprised. This is an exceedingly profound principle, although it is simple in concept, and takes the notion of information from the subjective plane to an objective level.

At an appointed time our knowledge of a situation can change dramatically based on "new" information. The difference in information before and after gives us the information content of the message giving us the outcome.

Let x be the probability that some event E will occur, where $0 \leq x \leq 1$. Assume further that a message is received that E has occurred. If $x=0.99$, you are not surprised and the message has little information content. On the other hand, if the value of x were 0.01, you are very surprised and the message would indeed have high information content. It would be reasonable to require that the information content $h(x)$ of a message be a decreasing function of the probability x. It is generally agreed to take the logarithm of the reciprocal of the probability x as the decreasing function, such that

$$h(x) = \log(1/x) = -\log x$$

In general, the base of the logarithm is open but base 2 is common in information theory and information can be measured in bits.

Example. Assume there are 2^N possible events and all have the same chance of occurrence 2^{-N}. The amount of information contained in the message that states the particular event that occurred is:

$$h(2^{-N}) = \log(1/2^{-N}) = N \text{ bits}$$

When the probability is not a power of two, then the information content will take on non-integer values.

Entropy

Consider a complete system $S = \{S_1, S_2, \ldots, S_n\}$ with corresponding probabilities $x_1, x_2, \ldots, x_n$, such that

$$\sum_{i=1}^{n} x_i = 1$$

where $0 \leq x_i \leq 1$ for $i=1,2,\ldots,n$.

When we receive a message that S_i has occurred, the information content of the message is $h(x_i)$=-$\log(1/x_i)$. Before the message is received, we do not know the event that will occur and the magnitude of the corresponding information content. On the other hand, we can develop the average information content (i.e, the *expected information content* $H(x)$) as:

$$H(x) = \sum_{i=1}^{n} x_i h(x_i) = \sum_{i=1}^{n} x_i \log(1/x_i) = -\sum_{i=1}^{n} x_i \log(x_i)$$

where

$$x_i \log(x_i)=0 \text{ if } x_i=0$$

The expected information content of a probability distribution is a measure of its disorder. In a distribution with n probabilities, the expected information approaches a maximum value as the n probabilities x_i approach $1/n$. The expected information as a measure of disorder is commonly known as *entropy* (Theil, 1967).

EXAMPLE. Consider the system $S = \{S_1, S_2\}$ with probabilities

$$p(S) = \{x_1, x_2\} = \{0.6, 0.4\}$$

The entropy of S is computed as

$$H(S) = \sum_{i=1}^{2} p(x_i) h(x_i)$$

$$= (0.6)\log(1/0.6) + (0.4)\log(1/0.4)$$
$$= 0.970951$$

The entropy is high because the possibilities approach being equally likely.

DEMPSTER-SHAFER THEORY

Dempster-Shafer theory (Shafer, 1976) has commanded a considerable amount of attention in the scientific and business communities, because it allows a knowledge source to assign a numerical measure to a proposition from a problem space, and provides a means for the measures accorded to independent knowledge sources to be combined. Dempster-Shafer theory is attractive because conflicting, as well as confirmatory, evidence from multiple sources may be combined.

The basis of Dempster-Shafer theory is the frame of discernment (Θ), introduced previously. Accordingly, a knowledge source may assign a numerical measure to a distinct element of Θ, which is equivalent to assigning a measure of belief to the corresponding proposition. In most cases, the numerical measure will be a basic probability assignment. A measure of belief may also be assigned to a subset of Θ or to Θ itself.

Support Functions

Consider a frame of discernment Θ and its power set denoted by 2^{Θ}. For example, given the frame:

$$\Theta = \{a, b, c\}$$

The power set is delineated as follows:

$$2^{\Theta} = \{\{a, b, c\},\ \{a, b\}, \{a, c\}, \{b, c\},\ \{a\}, \{b\}, \{c\}\}$$

In Dempster-Shafer theory, a knowledge source apportions a unit of belief to an element of 2^{Θ}. This belief can be regarded as a mass committed to a proposition and represents a judgment as to the strength of the evidence supporting that proposition. When viewed in this manner, evidence focuses on the set corresponding to a proposition; this set is called a *focal set.*

The support for a focal set is a function m that maps an element of 2^{Θ}, denoted by A, onto the interval $[0,1]$. Given a frame of discernment Θ and function $m: 2^{\Theta} \rightarrow [0,1]$, a support function is defined as follows:

$m(\phi) = 0$, where ϕ is the null set
$0 \leq m(A) \leq 1$, and

$$\sum_{A \subset 2^{\Theta}} m(A) = 1$$

The support function m is called a *basic probability assignment,* which is assigned by the knowledge source.

A support function is called a *simple support function* if it reflects, at most, one focal set not equal to Θ. A simple support function assigns a measure of belief to the focal set A, as follows:

$m(A) > 0$
$m(\Theta) = 1 - m(A)$
$m(B) = 0$, for all $B \subset 2^{\Theta}$ and $B \neq A$

The simple support function for a focal set A assigns a portion of the total belief exactly to A and not to its subsets or supersets. The remainder of the belief is assigned to Θ. Because certainty function must add up to 1, $m(\Theta) = 1 - m(A)$.

It is possible that a body of knowledge or evidence supports more than one proposition, as in the following case. If

$\Theta = \{a, b, c, d\}$
$A = \{a, b\}$

and

$B = \{a, c, d\}$

then the evidence supports two focal sets, which in the example, are A and B. If $m(A)=0.5$ and $m(B)=0.3$, then $m(\Theta)=0.2$.

A support function with more than one focal set is called a *separable support function*. Separable support functions are normally generated when simple support functions are combined.

The notion of combining simple support functions is a practical approach to the assessment of evidence. An analyst obtains information from a knowledge source, and it leads to an immediate conclusion – not with certainty, but with a certain level of belief. This is a normal straightforward means of handling human affairs and is precisely what people do. Then when additional information comes in, the various pieces of evidence are combined to obtain a composite picture of the situation.

Combination of Evidence

A method of combining evidence is known as Dempster's rule of combination (Dempster, 1967). Evidence would normally be combined when it is obtained from two different observations, each with the same frame of discernment. The combination rule computes a new support function reflecting the scope of the combined evidence.

If m_1 and m_2 denote two support functions, then their combination is denoted by $m_1 \oplus m_2$ and is called their *orthogonal sum*. The combination $m_1 \oplus m_2$ is computed from m_1 and m_2 by considering all products of the form $m_1(X) \bullet m_2(Y)$, where X and Y range over the subsets of Θ, and

$$\sum m_1(X) \bullet m_2(Y) = \sum m_1(X) \bullet \sum m_2(Y) = 1 \times 1 = 1$$

$m_1(X) \bullet m_2(Y)$ is the set intersection of X and Y.
For example, consider the frame of discernment

$\Theta = \{\text{healthy, test, sick}\}$

and focal sets A and B are based on two different observation over the same frame:

$A = \{\text{test, sick}\}$	$B = \{\text{healthy, tests}\}$
$m_1(A) = 0.8$	$m_2(B) = 0.6$
$m_1(\Theta) = 0.2$	$m_2(\Theta) = 0.4$

The combination rule is depicted via the following tableau:

	A	Θ
m_1	{test,sick}	{healthy,tests,sick}
m_2	0.8	0.2
B {healthy,tests} 0.6	{tests} 0.48	{healthy,tests} 0.12
Θ {healthy,test,sick} 0.4	{tests,sick} 0.32	{healthy,test,sick} 0.08

The entries are then combined as follows:

$m_1 \oplus m_2(\{\text{tests}\}) = 0.48$
$m_1 \oplus m_2(\{\text{healthy,tests}\}) = 0.12$
$m_1 \oplus m_2(\{\text{tests,sick}\}) = 0.32$
$m_1 \oplus m_2(\{\text{healthy,tests,sick}\}) = 0.08$
$m_1 \oplus m_2 = 0$ for all other subsets of Θ

Thus, for $A_i \cap B_j = A$ and $m_1 \oplus m_2 = m$, the combination rule is defined mathematically as:

$$m(A) = \sum_{A_i \cap B_j = A} m_1(A_i) \bullet m_2(B_j) / (1 - \sum_{A_i \cap B_j = \emptyset} m_1(A_i) \bullet m_2(B_j))$$

The denominator reflects a normalization process to insure that the pooled values sum to 1. So, in this instance, the normalization process yields the combination

$$A\oplus B = \{\{\text{healthy}\},0.63\},\{\{\text{tests}\},0.32\},\{\{\text{sick}\},0.05\}\}$$

after dividing the combined assessment by 0.38. Because the problem is well-structured, the representation can be simplified as

$$A\oplus B = \{0.63,0.32,0.05\}$$

For views $A=\{A_1,A_2,\ldots,A_n\}$ and $B=\{B_1,B_2,\ldots,B_n\}$, the combination rule can be simplified as

$$A\oplus B=\{A_1\times B_1/k,A_2\times B_2/k,\ldots,A_n\times B_n/k\}$$

$$\text{where } k=\sum_{i=1}^{n} A_i\times B_i$$

APPLICATIONS

Two applications of the preceding concepts are considered: a voting system and the elicitation of expert opinion.

Voting System

Consider a system in which a decision maker is required to vote on a well-structured issue, such as the selection of a candidate. Let the candidates be *Roberts, Richards,* and *Williams.* Assume further that the voter is influenced by three groups to cast a vote to their best interests. In a real sense, the voter is being pulled in three directions. Let the influential groups be *Party, Cause,* and *Lobby,* delineated as follows:

$$\text{Party} = \{\{\{\text{Roberts}\},0.6\},\{\{\text{Richards}\},0.3\},\{\{\text{Williams}\},0.1\}\}$$

Cause = {{{Roberts},0.4},{{Richards},0.2},{{Williams},0.4}}
Lobby = {{{Roberts},0.4},{{Richards},0.5},{{Williams},0.1}}

The information should be interpreted from a probabilistic view. For example, based on party affiliation, the voter will choose *Roberts* with a 0.6 probability, *Richards* with a 0.3 probability, and *Williams* with a 0.1 probability.

Clearly, the frame of discernment is {Roberts,Richards,Williams} and since the problem is well-structured, the basic probability assignments may be summarized as follows:

Party = {0.6,0.3,0.1}
Cause = {0.4,0.2,0.4}
Lobby = {0.4,0.5,0.1}

Table 1 summarizes the application of Dempster's rule of combination to this problem. First, *Party* is combined with *Cause* and then the result is combined with *Lobby* to obtain a composite picture of the result when the support functions are fused. In general, the evidence is complementary and this fact is reflected in the entropy that is reduced from the initial party affiliation of 1.29 to 0.965.

Support Function	bpa	Entropy
Party	{0.6,0.3,0.1}	1.29
Cause	{0.4,0.2,0.4}	1.52
Lobby	{0.4,0.5,0.1}	1.36
Party×Cause (=K)	{0.706,0.176,0.118}	1.16
K×Lobby	{0.739,0.230,0.031}	0.965

Table 1. Application of Dempster's Rule to the Voting System

Elicitation of Expert Opinion

Typically, experts do not agree, especially when system failure is concerned. Typical examples might be the crash of an expensive fighter aircraft or the collapse of a building. Consider a situation wherein the frame of discernment is $\{A,B,C\}$ reflecting that the failure could be caused by Component A, Component B, or Component C. Expert #1 believes the failure is due to Component A with probability 0.75, Component B with probability 0.15, or Component C with probability 0.10. Expert #2 believes the failure is due to Component A with probability 0.30, Component B with probability 0.20 or Component C with probability 0.50. The support function are:

Expert #1 = {{{A},0.75},{{B},0.15},{{C},0.10}}
Expert #2 = {{{A},0.30},{{B},0.20},{{C},0.50}}

The problem is well-structured so the probability assignments are summarized as follows:

Expert #1 = {0.75,0.15,0.10}
Expert #2 = {0.30,0.20,0.50}

Table 2 summarizes the application of Dempster's rule of combination to this problem. The opinion of the experts is summarized and reflects the differing opinions.

Support Function	bpa	Entropy
Expert #1 (=X)	{0.75,0.15,0.10}	1.05
Expert #2 (=Y)	{0.30,0.20,0.50}	1.49
X×Y	{0.634,0.084,0.282}	1.23

Table 2. Elicitation of Expert Opinion

The strong opinion of Expert #1 in favor of Component A, reflected in the low entropy, has a major influence on the consensus.

SUMMARY

The paper introduces the concepts of consensus, uncertainty, Dempster-Shafer theory, information theory, and entropy. The relationship between the combination of evidence and entropy is explored and examples from voting systems and the elicitation of expert opinion are given.

REFERENCES

Dempster, A.P. (1967), "Upper and Lower Probabilities Induced by a Multivalued Mapping," *The Annals of Statistics* 28:325-339.

Shafer, G. (1976), *A Mathematical Theory of Evidence*, Princeton, NJ: Princeton University Press

Theil, H. (1967), *Economics and Information Theory*, New York: American Elsevier Publishing Company, Inc.

END OF CHAPTER 7

8

TOWARD A UNIFIED ONTOLOGY OF TRUSTED IDENTITY IN CYBERSPACE

INTRODUCTION

The nation's digital infrastructure is in jeopardy because of inadequate provisions for privacy, identity, and security. The "everyone is free to do everything" mentally that would appear to be prevalent in America and worldwide has resulted in an onslaught of identity theft, fraud, digital crime, and an unnecessary concern over cyber security by many individuals. It is patently necessary for careful participants to operate defensively in cyberspace in order to protect themselves from the evils just mentioned. Those that do not act responsibly do so at their own peril. In fact, digital crime has served as a precursor to and is associated with physical crime. (OECD 2008)

The computer security and Internet communities have been generally responsive but apparently ineffective, so it is time for a third party to step in, take charge, and provide an infrastructure to assist in protecting the citizens of the world. (White House 2010) Similar concerns prevail in other developed countries, since many cybercrimes are perpetrated from lesser-developed countries without cyber awareness from legal, political, economic, and technical perspectives.

Moreover, there are no good reasons why underdeveloped countries should get a free-ride in this regard. The nations and organizations that lend a blind-eye to its residents that commit cybercrimes against persons and organizations in other countries will change their tune when cyber benefits are subsequently denied to them through political, social, and technological processes.

This paper is a contribution to the domain of ontological commitment as it applies to a description of subjects, objects, actions, and relationships as they pertain to the National Strategy of Trusted Identity in Cyberspace initiative.

BASIC CONCEPTS

Identity is a major issue in the security of modern information systems and the privacy of data stored in those systems. Identity and privacy concerns are commonly associated with behavioral tracking, personal-identifiable information (PII), the relevance of private data, data repurposing, and identity theft. (Windley 2005) We are going to approach the subject from an information systems perspective, recognizing that the inherent problems also apply to societal systems. Information systems are a good conceptual vehicle for the underlying security, identity, and privacy models, because data is typically stored off-premises and is under the control of a third-party service provider. When a third party gets your data, who knows what is going to happen to it? The main consideration may turn out to be a matter of control, because from an organizational perspective, control over information has historically been with the organization that creates or maintains it. From a personal perspective, on the other hand, persons should have the wherewithal to control their identity and the release of information about themselves, and in the latter case, a precise determination of to whom it is released and for what reason. Privacy issues are not fundamentally caused by technology, but they are exacerbated by employing the technology for economic benefit. After a brief review of identity and privacy to set the stage, we are going to cover identity theory, privacy theory, and identity requirements. This is a working paper on this important subject.

Identity

Identity is a means of denoting an entity in a particular namespace and is the basis of security and privacy – regardless if the context is digital identification or non-digital identification. We are going to refer to an identity object as a *subject*. A subject may have several identities and belong to more than one namespace. An identity denotation is based on attributes as suggested by Figure 1.

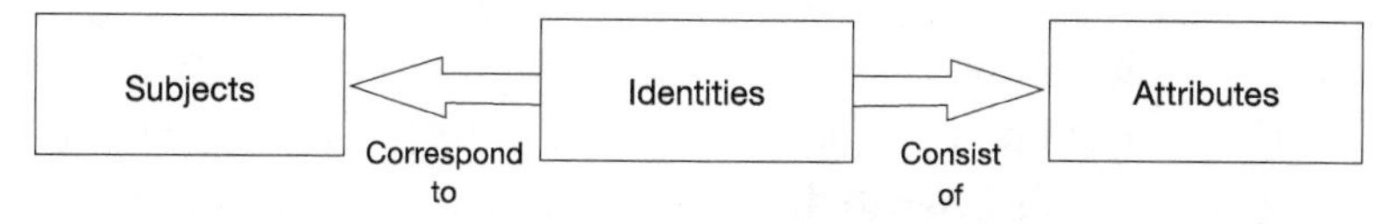

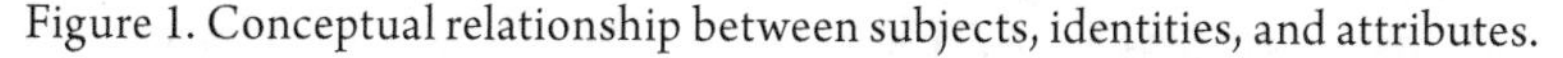

Figure 1. Conceptual relationship between subjects, identities, and attributes.

A pure identity denotation is independent of a specific context, and a federated identity reflects a process that is shared between identity management systems. When one identity management system accepts the certification of another, a phenomenon known as "trust" is established. The execution of trust is often facilitated by a third party that is acknowledged by both parties and serves as the basis of digital identity in information systems.

Access to computing facilities is achieved through a process known as authentication, whereby an entity makes a claim to its identity by presenting an identity symbol for verification and control. Authentication is usually paired with a related specification known as authorization to obtain the right to address a given service.

It is generally felt that a framework for understanding a technology should reflect the underlying concepts required for its development and subsequent acceptance as an operational modality. A technology should enable the delivery of value rather than constrain it, and that is our objective with this paper.

Privacy

Information systems typically process and store information about which privacy is of paramount concern. The main issue is identity, which serves as the basis of privacy or lack of it, and undermines the trust of individuals and organizations in other information-handling entities. The key consideration may turn out to be the integrity that organizations display when handling personal information and how accountable they are about their information practices. From an organizational perspective, control over information should remain with the end user or the data's creator with adequate controls over repurposing. From a personal perspective, the person should have the wherewithal to control his or her identity as well as the release of socially sensitive identity attributes. One of the beneficial aspects of the present concern over information privacy is that it places the person about whom data are recorded in proper perspective. Whereas such a person may be the object in an information system, he or she is regarded as the subject in privacy protection – as mentioned earlier. This usage of the word *subject* is intended to imply that a person should, in fact, have some control over the storage of personal information.

More specifically, the *subject* is the person, natural or legal, about whom data is stored. The *beneficial user* is the organization or individual for whom processing is performed, and the *agency* is the computing system in which the processing is performed and information is stored. In many cases, the beneficial user and the subject are members of the same organization.

The heart of the issue is *privacy protection*, which normally refers to the protection of rights of individuals. While the concept may also apply to groups of individuals, the individual aspect of the issue is that which raises questions of privacy and liberty

Privacy Assessment

The Federal Bureau of Investigation (U.S.A.) lists several criteria for evaluating privacy concerns for individuals and for designing computer applications: (FBI 2004)

- *What information is being collected?*
- *Why is the information being collected?*
- *What is the intended use of the information?*
- *With whom will the information be shared?*
- *What opportunities will individuals have to decline to provide information or to consent to particular uses of the information?*
- *How will the information be secure?*
- *Is this a system of records?*

Since privacy is a fundamental right in the United States, the above considerations obviously resulted from extant concerns by individuals and privacy rights groups. In a 2009 Legislative Primer, the following concerns are expressed by the Center for Digital Democracy: (CDD 2009, p. 2)

- Tracking people's every move online is an invasion of privacy.
- Online behavioral tracking and targeting can be used to take advantage of vulnerable consumers.
- Online behavioral tracking and targeting can be used to unfairly discriminate against consumers.
- Online behavioral profiles may be used for purposes beyond commercial purposes.

We are going to add to the list that the very fact that personal data is stored online is a matter of concern and should be given serious attention. Based on these issues, this paper is going to take a comprehensive look at the subject of identity in computer and human systems.

IDENTITY THEORY

The notion of identity is an important subject in philosophy, mathematics, and computer information systems. In its most general sense, identity refers to the set of characteristics that makes a subject definable. Each characteristic can be viewed as a single point in a three-dimensional Cartesian coordinate system where the axis are *subject, attribute,* and *value.* (Katzan 1975) Thus, the fact that George is twenty-five years old could be denoted by the triple <George, age, 25>. A set of characteristics over a given domain can uniquely identify a subject. This simple concept is the basis of privacy and identity in information systems and everyday life. The notion of identity applies to organizational subjects as well as to person subjects.

Knowledge, Attributes, and Identity

Identity is primarily used to establish a relationship between an attribute or set of attributes and a person, object, event, concept, or theory. The relationship can be direct, based on physical evidence, and in other cases, the relationship is indirect and based on a reference to other entities. In a similar vein, the relationship can be certain or uncertain, and in the latter case, based in deduction or inference. The relationship determines an element of knowledge. For example, the knowledge element "you are in your car" is a statement in which "you" and "your car" are things that exist and the "in" is a relationship. Direct knowledge is known by *acquaintance* and is evidenced by a physical connection. Indirect knowledge is determined through a reference to a particular with which the analyst is acquainted. The form is known as knowledge by *description.* (Russell 1912) *Direct knowledge* is determined through sense data, memory, or introspection. *Indirect knowledge* is determined through a reference to another particular, as in "the person who ran for Congress in 2004" or through a form of self-awareness where what goes on in subject's mind, for example, is estimated by an analyst's interpretation based on experience or self-evaluation.

Synthetic knowledge reflects certainty based on evidence inherent in the attribute values at hand. *Analytic knowledge* reflects a degree of uncertainty and is determined by deduction, as in "he is the only person with that 'attribute value'," or by inference based on known particulars, such as "all terrorists have beards." Inference, in this case, could be regarded as a form of derivative knowledge. The value of analytic knowledge is that it enables the analyst to exceed his or her limit of private experience.

Numerical and Qualitative Identity

Identity refers to the characteristics that make a subject the same or different. We are going to establish two forms of identity: numerical and qualitative. Two subjects are *numerically identical* if they are the same entity, such that there is only one instance. Two subject (or objects in this case) are *qualitatively identical* if they are copies or duplicates. In the popular movie *The Bourne Identity,* for example, the characters *Jason Bourne* and *David Web* are numerically identical, and the number of subjects is one. So it is with *Superman* and *Clark Kent* in another domain. On the other hand, a set of animals with the same biological characteristics – e.g., a species – are regarded as being qualitatively identical. The notion of qualitative identity is remarkably similar to the modern definition of a *category* informally defined as a collection of entities with the same characteristics, having the same values for the same attributes.

Theory of the Indiscernibles

An important aspect of identity theory is that subjects exhibit features of permanence and change, analogous to sameness and difference mentioned previously. We are going to discuss the concept of temporal identity in the next section. The notion of change implies that a subject undergoes transformation and also has a property that remains unchanged. Both Locke and Hume have proclaimed that change reflects the idea of unity and not of identity. Leibnitz proposed the *Theory of Indiscernibles* suggesting that subjects (i.e., objects or entities) that are indiscernible are identical. (Stroll 1967) The subject of indiscernibles has implications for information systems and attribute change. To what extent a change in a characteristic denotes a change in identity is an open item at this time and implies that there is a probabilistic aspect to identity.

Russell approaches the subject of identity from an alternate viewpoint, analogous to definite and indefinite articles. Russell proposes that a description may be of two sorts: definite and indefinite. A definite description is a name, and an indefinite description is a collection of objects x that have the property ø, such that the proposition øx is true. (Russell 1919) In the phrase *Dan Brown is a famous author*, for example, 'Dan Brown" is a name and the indefinite description is obvious, leading to the probabilistic link between a subject and a characteristic.

Temporal Identity

There is a rich quantity of philosophical literature on the change of identity over time. Are you the same person you were yesterday? Are there persistent attributes that allow for positive identity between time periods? As alluded to previously, entities in everyday life exhibit features of permanence and change. In the domain of personal identity, address attribute is a primary candidate for change. For example, John Smith lives at 123 Main Street. He moves out and another John Smith moves in. This is a distinct possibility in a crowded city. In there a concept in identity theory for this phenomena? Should an identity system take this eventuality into consideration?

There is a form of *attribute duality* between a person subject and an object subject. A subject – an object, such as a residence, in this case – is characterized by who lives there. For example, rich people live on Sutton Place in New York. The discussion leads to four related concepts: endurant identity, perdurant identity, endurant attribute, and perdurant attribute. Clearly, the term *endurant* refers to a noun that does not change, where perdurant refers to one that does. Thus, the identity problem is essentially translated to an operant problem of "recognizing identity."

PRIVACY THEORY

It has long been recognized that privacy is a two-edged sword, not only for individuals, but also for groups and organizations. Subjects have First and Fourth Amendment rights designed to protect against unwarranted disclosure of information with unlimited scope to unwanted parties without proper authorization by the subject. However, privacy considerations protect criminals and terrorists, in addition to ordinary citizens, groups, and organizations. Protections and other conventions used to safeguard trade secrets can also be employed to enable non-disclosure of design and manufacturing flaws from consumers and regulatory bodies.

Privacy has been in the news for at least forty years originating with Alan Westin's seminal book on the subject entitled *Privacy and Freedom,* published in 1967. Others have joined the struggle, namely (Westin 1977, Miller 1971, Katzan 1980, and Givens 2009) to reference only a few of many, with apologies to those not mentioned. One of the toughest problems facing the computer industry is data protection, summarized very well in 1971 by Arthur R. Miller: (Miller 1971, p.37)

> The new information technologies seem to have given birth to a new social virus – "data-mania." Its symptoms are shortness of breath and heart palpitations when contemplating a new computer application, a feeling of possessiveness about information and a deep resentment toward those who won't yield it, a delusion that all information handlers can walk on water, and a highly advanced case of

> astigmatism that prevents the affected victim from perceiving anything but the intrinsic value of data. Fortunately, only some members of the information-handling fraternity have been stricken by the disease.

This quote was written over 39 years ago; what would the author think about today's environment?

Privacy and Data Protection

Data protection is given the most attention when the privacy of an individual or an organization is jeopardized. According to Alan F. Westin: (Westin 1967)

> Privacy is the claim of individuals, groups, or institutions to determine for themselves when, how, and to what extent information about them is communicated to others.

Privacy is related to data protection, because it is an integral part of society and affects the behavior of its citizens. *Privacy is a service that a subject should expect from and be provided by society.* The physical state of being private has four primary attributes: solitude, intimacy, anonymity, and reserve, which supply group separation, group participation, group freedom, and personal protection, respectively. These states collectively provide the confidentiality required to participate in a civilized society. Concerns for privacy should be an integral part of a data protection program.

An organization requires privacy to achieve its basic objective – whether it is business, education, or government. The disclosure of private internal affairs affects "brand equity" and is detrimental to success.

Another consideration is personal surveillance – even though it may be socially or legally accepted. When a subject does not have control over its informational profile, there is no safeguard over its authenticity. Therefore, a double barreled approach, consisting of technology and regulation, is required for operating in a global economy. (Katzan, 1980, p. 44)

Information Control

Because of the widespread application of computer and communications technology, there has been a gradual trend among private institutions and government agencies to ignore the individual's need for privacy. Privacy safeguards are the individual's sole line of defense against the exercise of power through information control. Individuals can lose control of information about themselves in three ways:

1. Information obtained against the subject's wishes.
2. Information obtained from an agency against the wishes of the agency and of the subject.
3. Information willingly disclosed by the beneficial user or agency but against the subject's wishes.

Information obtained against a subject's wishes is an area in which privacy is normally expected. This category includes explicit attempts to obtain information and implicit methods where a subject is forced to disclose personal information. Typical actions are:

1. Searches and seizure
2. Compelled self-disclosure
3. Informers and secret agents
4. Participant monitoring
5. Public observation and recording of information
6. Consent for fear of reprisal
7. Disclosure for privilege

Some benefits are commonly associated with disclosure of private information, so the fine line between willing and unauthorized disclosure is frequently blurred. In the case of *Information obtained from an agency against the wishes of the agency and of the subject,* the conditions of privacy should apply to the agency as they do to the subject and are normally of concern because of computer security deficiencies and unauthorized access. In the case of *Information willingly disclosed by the beneficial user or agency but against the subject's wishes,* as in interagency transfers, accuracy and context are normally of concern. This is the prototypical *repurposing of information* that lies at the heart of most subjects' concerns over the disclosure of personal information.

Recordkeeping

Records typically fall into four classes: administrative, operational, intelligence, and statistical. In theory, *administrative records* are maintained by governmental agencies and give subjects their identity. For individuals, administrative records normally include birth certificates, diplomas, military discharge papers, driver's licenses, and immigration papers. For organizations, administrative records include certificates of incorporation and related documents. *Operational records* reflect tax and other certificates. *Intelligence records* are maintained by government agencies and represent security permissions and legal investigations. *Statistical records* can be obtained through an official questionnaire, as with the census, or from any of the other records that have been "cleansed" so as not to reflect personal information. Privacy safeguards are summarized in a far-reaching report by the Department of Health, Education, and Welfare (HEW 1973, pp. xx-xxi.):

1. There must be no personal data record-keeping systems whose very existence is secret.

2. There must be a way for an individual to find out what information about him is in a record and how it is used.
3. There must be a way for an individual to prevent information about him that was obtained for one purpose from being used or made available for other purposes without his consent.
4. There must be a way for an individual to correct or amend a record of identifiable information about him.
5. Any organization creating, maintaining, using, or disseminating records of identifiable personal data must assure the reliability of the data for their intended use and must take precautions to prevent misuse of the data.

The five principles are regarded as a Code of Fair Information Practice, emphasizing that privacy is a service that should be afforded to all citizens by other citizens, organizations, and the government in a free and open society.

Privacy Issues

The subject of privacy in all of its "multi-faceted dimensions" is of concern to many persons. Some individuals only wake up to the subject when their privacy is invaded and then quickly go back to sleep when the situation subsides, or they get tired of worrying about it. In the present context, Internet computing would seem to constitute a privacy threat to many persons and also organizations, because sensitive information is held by third-party service providers. However, having a third-party service provider is not a necessary condition for privacy invasion. The gang-of-three (government, employers, and education) would appear to be doing a good job with that. What are the specific issues about which we should be concerned? The topic has been addressed by the Privacy Rights Clearinghouse (PRC) in a document entitled "Privacy Today: A Review of Current Issues" developed by its director Dr. Beth Givens. (Givens 2009) The report lists twenty-three issues in privacy rights with a substantial description of each issue. The report highlights and summarizes the key issues and also contains links to special interest groups working on particular topics in that domain. We are going to concentrate on five subjects deemed relevant to the mission of this compendium:

- Biometrics
- Video surveillance and workplace monitoring
- Data profiling
- Behavioral tracking and targeting
- Records on the Internet

A selection from the PRC list is also necessary because every privacy subject has its privacy point and twenty-three primary issues are more than we can usefully cover in this paper. Here is a simple straight-forward case of an individual personal privacy concern. "Joe Smith is a good runner and ran a local marathon in 3 hours and 20 minutes. The marathon organizer lists the name, age, finishing time, finisher's place, and home city and state of all finishers of the race on the Web. Joe has two concerns. He is a bit embarrassed, because a couple of years ago, he ran the same race in less than 3 hours. So, in this case Joe would prefer not to have the results published online for everyone to see – that is, if anyone besides runners would be interested. Joe's friend Al has a different opinion. Al says, 'That is a great time Joe. My father, who is about your age, ran it in 3 hours and 10 minutes.' The second concern is more serious. Joe is 57 years old and is looking for a good position, since he was recently laid off. He is concerned that a prospective employer can Google him and determine his age from the online list of finishers, since age discrimination is a major concern for many employers in this country." If the race were run in Canada or Europe, on the other hand, the same information would not be available to outside persons, because of privacy laws.

We are going to present a descriptive technique that will apply the five selected dimensions, placing each dimension in a privacy-identity continuum.

Biometrics

The term *biometrics* refers to the use of bodily characteristics for identification, which can be exact or probabilistic. If you have been in the ROTC, the military, law enforcement, possess a government security clearance, or have been born recently, you have an exact biometric identity consisting of your set of fingerprints on file in an official place. A person's DNA and retinal scan are also supposedly exact biometric identifiers. Clearly, an exact biometric marking does in fact identify a particular individual. However, the assignment of a name from an appropriate namespace is quite another thing. If the task is to link an individual with a specific name, then there is some probability involved. The picture on an official passport, driver's license, or government issued identification is also regarded as an exact identifier. But, how exact is exact? As mentioned before, there is some risk in linking name identification between two or more types of identity.

Less exact biometrics, such as facial recognition, has been employed in social situations to identify persons of interest – such as at sporting events. Using facial geometry and other visual clues, facial recognition technology has been very successful in criminal investigation. But, what about the identification and recording of persons in a lawful demonstration, guaranteed as a First Amendment right? Everyone knows there are at least two kind of demonstrators: those persons participating in the physical part of a demonstration because they genuinely believe in the cause, and those persons with nothing else to do on a Saturday afternoon. As Dr. Givens writes, "As a result, innocent people can be wrongly identified as criminal (false-positives), and known criminal and suspected terrorists can fail to be detected altogether (false-negatives).

Video Surveillance and Workplace Monitoring

Low-cost video surveillance systems are prevalent in modern society, and their use ranges from convenience stores to day-care centers. In fact, video surveillance is so pervasive that most people think nothing about being under the eye of the camera. In criminal investigation, video surveillance is a useful identifier, albeit within some probabilistic limits, and also as an investigative tool.

Collectively, video surveillance and workplace monitoring can provide information related to the following phenomena:

- Facial recognition
- Unproductive employment activity
- Improper use of resources
- Violation of conditions of service

Use of an employer's computer or other resources is a good case in point. There are other forms of surveillance, such as Radio Frequency Identification (RFID) chips embedded in employee identification cards that can be used as an employee locator by recording when he or she leaves one room and enters another.

Keystroke monitors are sometimes used to determine ineffective use of equipment. Most employees do not seem to mind employee monitoring when on premises – but what can they do about it? Off premises and off hours surveillance and monitoring are quite another thing and exist as an open issue in privacy.

Data Profiling

Most of us are well represented in a multitude of gang-of-three databases, such as the tax bureau, social security administration, state motor vehicles office, education records, employment files, insurance, and health records. Information of this type can be regarded as the operational part of the fabric of life. We can temper the intrusion but not totally eliminate it, because it is paramount to identity determination and service management. Identities are linked by numbers, such as the social-security number, name, date of birth, telephone number, address and ZIP code, mother's maiden name, and even mother's birthday. It is even possible to find the social security number of an unrelated deceased person on the Web. Immigration records are also easily obtainable. The Privacy Act of 1974 and its amendments generally cover governmental data protection and profiling.

There is another form of data that is involuntarily collected about individuals where there is some choice involved, such as personal expenditures, lifestyle, Internet activity, political activity and donations, and so forth. Supermarkets, bookstores, department stores, health stores, fitness centers, libraries, toll booths, big-ticket retailers, travel agencies, magazine publishers, and airlines – all contain personal data on individuals. An idea of interests, activities, and expenditures are available from credit card purchases, bank records, and operational files of business, governmental, and educational institutions. Thus, it is quite easy for an interested party to create a *data profile* of a person.

Pundits claim that profile data determines who or what we are. However, there is a tendency to interpret data based on the psychological perspective of the profiler. If you subscribe to "guns and ammo," does that indicate that you are a terrorist, member of the local shooting club, an Olympic athlete, worker in a sporting goods store that sells guns, or a medical professional who uses a service to provide magazines for the waiting room.

It has been reported that search providers turn over search queries of individuals to the agencies of the government. (Conti 2009, pp.259-298) This is a modern form of data profiling. A method, termed *chaffing,* is mentioned to widen the search domain and provide some protection. So, if you are going to search for a controversial person, you might also want to search for some non-threatening person to widen the search area.

Behavioral Tracking and Targeting

`Behavioral tracking and targeting is an area of privacy concern related to data profiling with emphasis on what a subject does. Here is a typical scenario. A subject rents a car and drives that vehicle out of state or out of the country by accident or by intention. When the car is turned in to the agency, the renter is charged an enormous penalty. The fine print in the contract was not read, because the renter is usually out of his or her element or just in a hurry. How did the agency know of the unfortunate travel? The car rental agency used a global positioning system (GPS) device to track the path of the vehicle. In addition to GPS tracking, license plate tracking, implemented through highway cameras, is also widely used by state and local law enforcement officials. There is always a stated reason why organizations do things, but in the case of privacy, the main problem is the repurposing of collected data. Through data mining technology, computers can identify patterns based on happenstance, rather than purposeful activity. Here is another example: At the time this paragraph was written, the state of Arizona decided to take border control into its own hands. The federal government could do it and can do it, but we live in a large country with enough problems to go around. Getting the right person's or an organization's attention at the right time usually takes some up front planning. Demonstrations ensue for varying reasons, including the possibility that certain outside people want to stir up trouble. Proper officials are looking into persons flying into Arizona with recently booked tickets for travel lasting only a few days and are doing some data mining to identify those persons. Are identified persons demonstration instigators or grandparents attending a graduation ceremony. Regardless, they are prime candidates for behavioral tracking. In an era of supercomputing, piecing together a travel itinerary is not a major task. All that is needed is a subject to track.

The subject of behavioral tracking also includes the practice of collecting and compiling consumers' online activities, interests, preferences, and/or communications over time. (Givens 2009, 18 of 23) This form of behavioral targeting serves as the basis for advertising and other forms of marketing. Web browsing is a primary source of information in this regard.

There is also a growing trend by Internet service providers (ISPs) to use deep packet investigation (DPI) to look at email, Web sites visited, music, video sharing, and downloads by inspecting the data packets that constitute Internet traffic. This form of privacy intrusion is a major challenge to privacy advocates.

Records on the Internet

There is a tendency in society for persons in a political or geographical jurisdiction to be generally the same. This refers to attitudes, culture, psychological properties, and so forth. Between countries, however, there tend to be some differences between the two groups of people. People from Switzerland are different from people from England. The same idea holds true for people from Minnesota and Georgia, for example. We are referring to what is acceptable behavior, from a cultural viewpoint, or "would you like that person living next door."

The disclosure of public records in an open government is not sensitive to cultural differences, since the context for the information in government-managed files does not travel with the information. Citizens in one area may be more or less sensitive to the content of public information than persons from another – especially in a large country. The "one size fits all" mentality of public disclosure is a subject that frustrates privacy advocates.

Nevertheless, divorce records, criminal records, under-age convictions, bankruptcy proceedings, DIU convictions, motor vehicle records, and so forth, are all publically available through mailing list and information brokers. All an identity thief or stalker needs is a Social Security number and $19.95. The motivation for many, if not most, automobile breakins in modern times is an attempt to obtain personal information in the glove box, even thought the thief may also take a camera from the rear seat.

IDENTITY REQUIREMENTS

It would appear that there are two essential problems in identity theory: protection of identity and recognition of identity. Protection refers to the safeguarding of one's identity from unwanted intrusion into personal affairs, and is reflected in the identity principles that follow. Recognition refers to the use of identity measures to classify certain persons, based on the combination of evidence and abductive inference. This characterization of the identity problem reflects two edges of the same sword.

Identity Principles

It is generally regarded that effective identity governance should be based on a set of principles to guide the professional activities of IT managers, security officers, privacy officers, and risk management. (Salido 2010, OECD 2010) As delineated, the principles would be based on efficacy in governance, risk management, and compliance with the following objectives:

> **Governance.** Assurance that the organization focuses on basic issues and who is responsible for actions and outcomes.
>
> **Risk Management.** Assurance that procedures are in place for identifying, analyzing, evaluating, remdying, and monitoring risk.
>
> **Compliance.** Assurance that actions are within the scope of social and legal provisions.

In accordance with the stated objectives, we can delineate the eight core principles of effective and efficient identity management. (OECD op cit., p.3)

> Principle #1. Collection Limitation Principle – there should be prudent limits on the collection of personal data with the knowledge or consent of the subject.
>
> Principle #2. Data Quality Principle – personal data should be relevant to stated purposes and be accurate, complete, and up-to-date.
>
> Principle #3. Purpose Specification Principle – the purpose of the data collection should be specified beforehand.
>
> Principle #4. Use Limitation Principle – data should be used only for the use specified and not be repurposed.
>
> Principle #5. Security Safeguards Principle – personal data should be safeguarded by reasonable and state-of-the art security facilities.
>
> Principle #6. Openness Principle – the technical infrastructure for protecting personal data should be open as to development, practices, and policies.
>
> Principle #7. Individual Participation Principle – the subject should have the right to definitive information concerning the personal data collected, methods used, and safeguards employed and have the right to challenge the procedures employed.

> Principle #8. Accountability Principle – social, business, educational, and governmental data controllers should be required by legal or regularity means to abide by principles 1-8 and be accountable for violations of their provisions.

The eight principles of identity agree in part and parcel to Cavoukian's "7 Laws of Identity, listed as follows: personal control and consent; minimal disclosure for limited use; need to know access; user-directed identity; universal monitoring of the use of identification technology; human understanding and involvement; and consistent access and interface to personal data. (Cavoukian 2010)

Identity Analytics

An important aspect of identity theory concerns whether a certain subject is a member of a group of interest. The basis for this form of identity determination is that identity is a function of the subject's namespace and attributes. A subject belongs to a category if it possesses the attributes that define the category. Another approach is to employ a knowledge source to determine a subject's group membership. This is the method we are going to use in this section. A popular characterization of the problem would be, "Is suspect A a member of group T?" or in short form, "Is A a T?" Clearly, the methods would apply to most diagnostic systems, such as medical diagnosis, auto repair, and the analysis of aircraft failures. We are going to propose two methods of analysis: the combination of evidence (Shafer 1976, Katzan 2006, Katzan 2010) and abductive inference (Josephson 1996).

With the *combination of evidence,* a certain level of belief is afforded a knowledge source, as in the following scenario:

> *We are trying to identify subjects that belong to a certain group G. We know about the group G and its attributes. We have a paid knowledge source K_1 that informs us that subject A is a member of G. However, K_1 is not always correct, and we know that. We have used K_1 enough to know that he provides us with information when he needs money. We have an intuitive belief of how often he is correct. Fortunately, we have another source K_2 that can supply similar information. K_2 is not as hungry for money as K_1, and his opinion frequently runs contrary to K_1's. We would like to use analytics to combine the information from K_1 and K_2 so as to obtain a composite picture of the situation.*

The relations between the knowledge sources and the subject are represented by the following mappings:

$$K_1 \rightarrow A$$
$$K_2 \rightarrow A$$

and the characteristics of the relationships are given as:

$$A = \{m, n\}$$
$$K_1 = \{r, u\}$$
$$K_2 = \{c, i\}$$

The question is whether A is a member of G, denoted by m, or not a member of G, denoted by n. As far as K_1 is concerned, he might be telling us what he thinks we want to hear, so his judgment is classed as reliable, denoted by r, or unreliable, denoted by u. K_2 is simply correct or incorrect, denoted by c or i, respectively. Through a method known as belief propagation (Katzan 2010a), the knowledge is transferred from the problem space to the solution space, resulting in the following representation:

Source	*Representation*
K_1	{[(m), *p*]. [(m, n), 1-*p*]}
K_2	{[(n), *q*]. [(m, n), 1-*q*]}

The results of belief propagation assign the mass (p) of the information received from K_1 to (m) and the remainder of the belief is assigned to (m, n). A similar argument applies to K_2 such that the mass (q) of that belief is assigned to (n) and the remainder to (m, n). Using Dempster's rules of combination (Dempster 1967), the resulting forms can be combined yielding the following assessment in the solution space:

$$\left[(m), \frac{p(1-q)}{1-pq}\right], \left[(n), \frac{(1-p)q}{1-pq}\right]. \left[(m,n), \frac{(1-p)(1-q)}{1-pq}\right]$$

using symbolic math from calculations in *Mathematica*™. Applying the expression to several values of p and q yields the following results:

K_1 (p)	K_2 (q)	$K_1 \oplus K_2$
.6	.7	{[(m), 0.310], [(n), 0.483], [(m, n), 0.207]}
.7	.5	{[(m), 0.538], [(n), 0.231], [(m, n), 0.231]}

This is what we wanted to show. QED.

An alternate methodology, known as *abductive inference,* is used to determine the probable cause of group membership. (Josephson 1996) As with many forms of diagnosis, we have an event or condition and wish to determine the probable cause of the occurrence. A person may have a condition, such as liver disease, or a physical system may fail, such as a fighter aircraft. The list of probable causes in each instance is called the *differential*. Abductive inference is often referred to as, "turning modus ponens induction on its head." Abduction takes the following pattern:

> E is an event or a collection of data
> C explains E
> No other hypothesis can explain E and well as C
> --
> Therefore, C is probably true

For example, J attends a training camp that is associated with membership in a militant group. There could be several possible causes for this phenomenon, two of which are that J is a militant or wants to be one. J could also be a journalist wanting to find out about things, but that is definitely less probable. The use of subjective probabilities, assigned through abductive inference, can be an analytic technique in its own right, or it can be used as input to a "combination of evidence" methodology.

Abductive inference and consensus theory would appear to be promising research topics for the modern world.

SUMMARY

Personal identity and privacy are important topics in the modern world of communications and the Internet. Most citizens are not aware of the major issues or do not realize the serious nature of identity theft and privacy invasion. The academic community is needed to foster attention to this subject, and this paper attempts to spotlight the major concerns. Accordingly, this paper is an admixture of topics that include identity, identity theory, privacy, and privacy theory, along with a summary of the major aspects of each domain.

There is due consideration given to identity requirements through a set of identity principles and some proposals for identity analytics. In the latter instance, evidential methods are presented as promising research topics.

REFERENCES

1 ACLU of Northern California. 2010. *Cloud Computing: Storm Warning for Privacy?* www.dotrights.org, (downloaded 3/11/2010).

2 Black, M. 1952. Identity of Indiscernibles. *Mind* 61:153. (Secondary reference.)

3 Cavoukian, A. 2009. *Privacy in the Clouds.* Toronto: Information and Privacy Commission of Ontario (www.ipc.on.ca).

4 Cavoukian, A. 2010. 7 Laws of Identity: The Case for Privacy-Embedded Laws of Identity I the Digital Age." Toronto: Information and Privacy Commission of Ontario (www.ipc.on.ca).

5 Katzan, H. 2010. *Privacy, Identity, and Cloud Computing,* New York: iUniverse, Inc.

6 OECD 2010. OECD Guidelines on the Protection of Privacy and Transborder Flows of Personal Data. www.oecd.org. (downloaded 3/23/2010).

7 Salido, J. and P. Voon. 2010. A Guide to Data Governance for Privacy, Confidentiality, and Compliance: Part 1. The Case for Data Governance. Microsoft Corporation.

8 Windley, P. 2005. *Digital Identity,* Sebastopol: O'Reilly Media, Inc.

END OF CHAPTER 8

9

PRODUCT ANALYTICS BASED ON DEMOGRAPHIC DEMOCRATIZATION

INTRODUCTION

The efficacy of decisions made under uncertainty on product allocation and selection is dependent upon two important concepts: the representation of the problem domain and the completeness of the solution space. A *category* is a means of representing the problem domain so that relevant structural information, based on demographics and consumer preferences, can be determined, resulting in improved decision-making. A *frame of discernment* (Shafer 1976; Katzan 1992, 2006, 2008) is a set of mutually exclusive and collectively exhaustive possibilities for the solution space. We are going to apply analytics to the task of product determination as a means of reducing the risk inherent in conventional statistical methods.

Demographics and Preferences

The selection of a product mix based on client preferences is an exceedingly complex task, because of the combinatorics of the independent variables. In an automobile selection process, for example, the number of consumer categories, such as gender, age, and education, is reasonably large, and the structural elements within each category are numerous enough to yield a large number of combinations. With the following categories, for example:

C = {gender, age, education}
gender = {male, female}
age = {<=25, 26-55, >=56}
education = {none, attended, grad}

the number of combinations of structural elements is $2 \times 3 \times 3$ or 18. In a typical set of eight demographic categories, the number of combinations is considerably greater than 2^8.

We propose a methodology based on categorical analytics, and the use of consensus theory (Katzan 2006) for combining information. The steps that comprise product analytics are:

1. Compute analytic alternatives for each category based on historical data. With products A and B for the gender category the process could yield the following data snippet:

Category	*Structural Element*	*Structural Probability*	*Preference A*	*Preference B*
gender	male	0.4	0.7	0.3
	female	0.6	0.5	0.5

2. Compute composite categorical probabilities by combining the analytic alternatives with preferences, as follows:

Category	*Probability A*	*Probability B*
gender	0.58	0.42

where the value 0.58 is computed as $0.4 \times 0.7 + 0.6 \times 0.5$, and the value 0.42 is computed as $0.4 \times 0.3 + 0.6 \times 0.5$.

3. Combine the resulting set of composite categorical probabilities, such as

Category	*Probability A*	*Probability B*
gender	0.58	0.42
age	0.44	0.56
education	0.46	0.54

using consensus theory yielding a set of probabilities for the solution set from which a decision maker can establish a product mix based on posterior probabilities concerning the client is most likely to purchase.

We call this process the **democratization** of product offerings based on preference data, because previous customers are essentially voting on what products to offer.

Category

A category is a means of structuring a problem domain with the objective of engaging in a predictive modality in which one or more future states may be identified and analyzed. Let $\mathbf{C}_i$ be one of the categories used to stratify the problem domain such that the collection

$$\mathbf{C} = \{\mathbf{C}_1, \mathbf{C}_2, \ldots, \mathbf{C}_n\}$$

represents a complete conceptualization of the dynamics under investigation and n is the number of categories.

Associated with each category is a set of probabilities representing an assessment of a future outcome based on its underlying categorical imperative. Thus, a category is a mechanism for isolating a single view of the problem under consideration. The ontological definition of a *category,* as a conceptual entity with no attributes in common with other categories, is adopted in this paper. The mutually exclusive set of possibilities under investigation is known as the frame of discernment. It is covered next, followed by a presentation of an approach to the structural analysis of categories. Finally, a product selection application is used as a demonstrative example that gives some insight into how the methods can be applied to other problems.

Frame of Discernment

A frame of discernment is a means of representing the possibilities under consideration, as in the following examples:

P = {sedan, wagon, suv}
E = {McCain, Obama}

Clearly, the elements in a frame of discernment are, in fact, propositions that can be interpreted as events or states. Thus, if component s_i of system *S* over domain *V* were associated with the symbol "sedan," then that state is equivalent to the proposition, "The true value of *V* for component s_i is sedan," or in ordinary language, "s_i prefers sedan."

Accordingly, the set S of propositions S_i,

$$S = \{S_1, S_2, \ldots, S_n\}$$

represents the collection of states of a system under analysis. Clearly, at an agreed upon point in time, one proposition is true and the others are false.

Uncertainty

Prior to the agreed point in time (τ), we obviously do not know the state of the system under analysis or its components with any degree of certainty. The expectation that a part of the system will be in a particular state at time τ is denoted by a real number $p(S_i)$ associated with each of the propositions in the frame $S = \{S_i\}$, i=1,2,…,n, such that

$$0 \leq p(S_i) \leq 1$$

and

$$\sum_{i=1}^{n} p(S_i) = 1$$

This is simply the addition rule for mutually exclusive events.

CONSENSUS THEORY

Consensus theory is a methodology for combining evidence based on Dempster-Shafer theory (Shafer 1976; Katzan 1992, 2006) and the mathematical combination of evidence (Dempster 1967). Dempster-Shafer theory has commanded a considerable amount of attention in the scientific and business communities, because it allows a knowledge source to assign a numerical measure to a proposition from a problem space and provides a means for the measures accorded to independent knowledge sources to be combined. Dempster-Shafer theory is attractive because conflicting, as well as confirmatory, evidence from multiple sources may be combined.

The basis of Dempster-Shafer theory is the frame of discernment (Θ), introduced previously. Accordingly, a knowledge source may assign a numerical measure to a distinct element of Θ, which is equivalent to assigning a measure of belief to the corresponding proposition. In most cases, the numerical measure will be a basic probability assignment. A measure of belief may also be assigned to a subset of Θ or to Θ itself.

Support Functions

Consider a frame of discernment Θ and its power set denoted by 2^{Θ}. For example, given the frame:

$$\Theta = \{a, b, c\}$$

The power set is delineated as follows:

$$2^{\Theta} = \{\{a, b, c\}, \{a, b\}, \{a, c\}, \{b, c\}, \{a\}, \{b\}, \{c\}\}$$

In Dempster-Shafer theory, a knowledge source apportions a unit of belief to an element of 2^{Θ}. This belief can be regarded as a mass committed to a proposition and represents a judgment as to the strength of the evidence supporting that proposition. When viewed in this manner, evidence focuses on the set corresponding to a proposition; this set is called a *focal set.*

The support for a focal set is a function m that maps an element of 2^{Θ}, denoted by A, onto the interval $[0,1]$. Given a frame of discernment Θ and function $m: 2^{\Theta} \rightarrow [0,1]$, a support function is defined as follows:

$m(\phi) = 0$, where f is the null set
$0 \leq m(A) \leq 1$, and

$$\sum_{A \subset 2^{\Theta}} m(A) = 1$$

The support function m is called a *basic probability assignment,* which is assigned by the knowledge engineer or domain specialist.

A support function is called a *simple support function* if it reflects, at most, one focal set not equal to Θ. A simple support function assigns a measure of belief to the focal set A, as follows:

$m(A)>0$
$m(\Theta)=1-m(A)$
$m(B)=0$, for all $B \subset 2^{\Theta}$ and $B \neq A$

The simple support function for a focal set A assigns a portion of the total belief exactly to A and not to its subsets or supersets. The remainder of the belief is assigned to Θ. Because certainty function must add up to 1, $m(\Theta)=1-m(A)$.

It is possible that a body of knowledge or evidence supports more than one proposition, as in the following case. If

$\Theta = \{a, b, c, d\}$
$A = \{a, b\}$

and

$B = \{a, c, d\}$

then the evidence supports two focal sets, which in the example, are A and B. If $m(A)=0.5$ and $m(B)=0.3$, then $m(\Theta)=0.2$. A support function with more than one focal set is called a *separable support function*. Separable support functions are normally generated when simple support functions are combined.

The notion of combining simple support functions is a practical approach to the assessment of evidence. An analyst obtains information from a knowledge source, and it leads to an immediate conclusion – not with certainty, but with a certain level of belief. This is a straightforward means of handling human affairs and is precisely what people do. Then, when additional information comes in, the various pieces of evidence are combined to obtain a composite picture of the situation.

Combination of Evidence

A method of combining evidence is known as Dempster's rule of combination ([Dem67]). Evidence would normally be combined when it is obtained from two different observations, each over the same frame of discernment. The combination rule computes a new support function reflecting the consensus of the combined evidence.

If m_1 and m_2 denote two support functions, then their combination is denoted by $m_1 \oplus m_2$ and is called their *orthogonal sum*. The combination $m_1 \oplus m_2$ is computed from m_1 and m_2 by considering all products of the form $m_1(X) \bullet m_2(Y)$, where X and Y range over the elements of Θ; $m_1(X) \bullet m_2(Y)$ is the set intersection of X and Y combined with the product of the corresponding probabilities.

For example, consider the frame of discernment

$$\Theta = \{\text{healthy, tests, sick}\}$$

and views A and B, based on two different observation over the same frame:

$$X = \{\{\text{healthy}\},0.6\},\{\{\text{tests}\},0.3\},\{\{\text{sick}\},0.1\}\}$$
$$Y = \{\{\text{healthy}\},0.4\},\{\{\text{tests}\},0.4\},\{\{\text{sick}\},0.2\}\}$$

The entries are combined as follows using Dempster's rule of combination:

$$m_1 \oplus m_2(\{\text{healthy}\}) = 0.24$$
$$m_1 \oplus m_2(\{\text{tests}\}) = 0.12$$
$$m_1 \oplus m_2(\{\text{sick}\}) = 0.02$$
$$m_1 \oplus m_2(\{\emptyset\}) = 0.62$$

Thus, for $A_i \cap B_j = \mathbf{A}$ and $m_1 \oplus m_2 = \boldsymbol{m}$, the combination rule is defined mathematically as:

$$\boldsymbol{m}(\mathbf{A}) = \sum_{A_i \cap B_j = \mathbf{A}} m_1(A_i) \bullet m_2(B_j) / (1 - \sum_{A_i \cap B_j = \emptyset} m_1(A_i) \bullet m_2(B_j))$$

The denominator reflects a normalization process to insure that the pooled values sum to 1. So, in this instance, the normalization process yields the combination

$$X \oplus Y = \{\{\text{healthy}\},0.63\},\{\{\text{tests}\},0.32\},\{\{\text{sick}\},0.05\}\}$$

after normalization by dividing the combined assessment by (1-0.62) or 0.38. Because the problem is well-structured, the representation can be simplified as

$$X \oplus Y = \{0.63,0.32,0.05\}$$

For views $A=\{A_1,A_2,\ldots,A_n\}$ and $B=\{B_1,B_2,\ldots,B_n\}$, the combination rule can be simplified as

$$A \oplus B = \{A_1 \times B_1/k, A_2 \times B_2/k, \ldots, A_n \times B_n/k\} \qquad [1]$$

where $k = \sum_{i=1}^{n} A_i \times B_i$

We will refer to equation [**1**] as the *simplification rule.*

An example of the preceding concepts is demonstrated through the elicitation of expert opinion.

Elicitation of Expert Opinion

Typically, experts do not agree, especially when system failure is concerned. A typical example might be the crash of an expensive fighter aircraft or the collapse of a building. Consider a situation where the frame of discernment is $\{A,B,C\}$ denoting that the failure could be caused by Component A, Component B, or Component C. Expert #1 believes the failure is due to Component A with probability 0.75, Component B with probability 0.15, or Component C with probability 0.10. Expert #2 believes the failure is due to Component A with probability 0.30, Component B with probability 0.20, or Component C with probability 0.50. The support function are:

Expert #1 = {{{A},0.75}, {{B},0.15}, {{C},0.10}} = {0.75, 0.15, 0.10}
Expert #2 = {{{A},0.30}, {{B},0.20}, {{C},0.50}} = {0.30, 0.20, 0.50}

Table 1 summarizes the application of the simplification rule to this problem. The opinion of the experts is summarized and reflects the differing opinions.

Support Function	Probability Assignment	Entropy
Expert #1 (=X)	{0.75, 0.15, 0.10}	1.05
Expert #2 (=Y)	{0.30, 0.20, 0.50}	1.49
X×Y	{0.738, 0.098, 0.164}	1.08

Table 1. Elicitation of Expert Opinion

The strong opinion of Expert #1 in favor of Component A, reflected in the low entropy, has a major influence on the consensus.

STRUCTURAL ANALYTICS

A problem domain is composed of categories, each of which is defined by a set of alternate structures. In a product example, for example, the category gender could be defined as

gender = {male, female}

based on a structural assessment, such as demographics. In this instance, the category gender is one of many viewpoints of an underlying decision situation, which could be a vote in an election or a position on an important issue. We are going to argue that in many unstructured decision-making problems, the probabilistic outcome can be based on structural, rather than, preferential elements. **What makes an unstructured decision so complex is that there are usually several categories "tugging at the decision maker."** We are going to show how categorical assessments can be combined to form a composite assessment of a decision under consideration. Through the technique known as structural analysis, we are going to assign probabilities to the elements of the frame of discernment from a given category, and then use consensus theory to combine the probabilities from the various categories. For example, a choice based on gender could go one way and a choice based on age could go another way. A realistic assessment would involve the combination of the two factors.

Structural Elements

Each category $\mathbf{C}_i$ is comprised of a set of structural elements $\mathbf{S}_i = \{S_{i1}, S_{i2}, S_{i3}, \ldots, S_{ik}\}$, where k is the number of structural elements in category $\mathbf{C}_i$. Consider the previously given universe defined as:

$C = \{C_1, C_2, C_3\}$ = {gender, age, education}, where
$\mathbf{S}_1$ = gender = {male, female}, and
$\mathbf{S}_2$ = age = {<=25, 26-55, >=56}
$\mathbf{S}_3$ = education = {none, attended, grad}

where $\mathbf{S}_1$, $\mathbf{S}_2$, and $\mathbf{S}_3$ are defined respectively as

S_{11} = male	S_{21} = <=25	S_{31} = none
S_{12} = female	S_{22} = 26-55	S_{32} = attended
and k = 2	S_{23} = >=56	S_{33} = grad
and k = 3	and k = 3	

Each problem domain is represented by a set of categories, each of which is a special lens into the underlying problem. Each category is defined as a set of structural components that define it. The categorical demographics in an election, for example, could be party, gender, age, and so forth. In the immediate example, the categories are gender, age, and education.

Structural Probabilities

Each structural element has a demographic probability $p(S_{ij})$

where $\sum_{j=1}^{k} p(S_{ij}) = 1$ for category i and structural element j in category i
and k is the cardinality of $\mathbf{S}_{i.}$

Accordingly, for category $\mathbf{C}_i$ and its structure $\mathbf{S}_i$, the probability set would be expressed as:

$$\mathbf{P}_i = \{p(S_{i1}), p(S_{i2}), \ldots, p(S_{ik})\}$$

For example, the probability set for category #1 (gender), could be

$$\mathbf{P}_1 = \{0.4, 0.6\}$$

representing male and female.

Each $\mathbf{P}_i$ represents the "probability of occurrence" in the universe of study of the structural elements of category i. This is demographic information.

Analytic Alternatives

In this form of analysis, each category $\mathbf{C}_i$ has an associated probability set $\mathbf{P}_i$. Each structural element has a corresponding probability $p(S_{ij})$ in $\mathbf{P}_i$. That probability represents the likelihood that an object selected at random from category $\mathbf{C}_i$ would be S_{ij}. Another interpretation is that a value in $\mathbf{P}_i$ gives the proportion of the corresponding structural element in $\mathbf{C}_i$. The following table gives another example:

Category	*Structural Element*	*Probability*
age	<25	0.4
age	26-55	0.4
age	>=56	0.2

The structural probabilities, alternately regarded as structural proportions, give a means of describing the environment in which a decision is to take place. In a product analysis, the environment would be the consumer demographics.

Preference Set

Each structural element is assigned a *preference set* over the frame of discernment from a knowledge source, such as a poll, survey, or statistical data. The probabilities in the preference set are the decision variables. For example, we might know that male person prefers product A with probability 0.7 and product B with probability 0.3. The set {0.7, 0.3} is known as the *preference set.*

Thus, for each structural element S_{ij} for all categories, there exists a preference set

$ps(S_{ij}) = \{p_{ij}(\Theta^1), p_{ij}(\Theta^2), \ldots, p_{ij}(\Theta^t)\}$, where t is the cardinality of the frame of discernment,

and

$$\mathbf{\Theta}^k = \{\Theta^1, \Theta^2, \ldots, \Theta^t\}$$

Clearly, $\sum_{k=1}^{t} p_{ij}(\Theta^k) = 1$ for all i and j.

Composite Probabilities

Composite categorical probabilities for each element in the frame of discernment are computed by combining the structural probabilities and corresponding preference set as follows:

$$P(\Theta_{it}) = \sum_{j=1}^{t} ((p(S_{ij}) \cdot ps(S_{ij})) \qquad [2]$$

where the index i runs through the categories and the index t runs through the alternatives in the frame of discernment.

Categorical Probabilities

The composite probabilities represent a summation of the preference for each element of the frame of discernment for each category. The result is a set of independent categorical assessments of the problem domain from different viewpoints represented as probabilities, as follows:

$$\mathbf{C}_i = \{P(\Theta_{i1}), P(\Theta_{i2}), \ldots, P(\Theta_{it})\}$$

where t is the cardinality of the frame of discernment, as defined previously.

Using the simplification rule [**1**], we derive a combined assessment of categories $\mathbf{C}_i$ and $\mathbf{C}_j$ of the form

$$\mathbf{C}_i \oplus \mathbf{C}_j$$

So that if

$$\mathbf{C}_1 = \{0.54, 0.46\} \text{ and } \mathbf{C}_2 = \{0.58, 0.42\}$$

Then

$$\mathbf{K} = \mathbf{C}_i \oplus \mathbf{C}_j = \{0.62, 0.38\}$$

The evidence is complementary, and that fact is reflected in the combined assessment.

PRODUCT ALLOCATION APPLICATION

One of the most familiar unstructured decision applications is the prior assessment of the products that customers will purchase. The major determinants of what products people will purchase can be combined into four well-known categories: gender, age, education, and race. The structural elements for each of the categories are given in Sheet 1, along with the respective structural probabilities. The columns are titled "Demographics." For the category gender, the structural element male has a probability (or proportion), for example, of 0.4. Associated with each structural element is a preference set for that element over the frame of discernment, which is {A, B}. In this case, a person in gender/male, would choose product A with probability 0.7 and B with probability 0.3.

Categorical probabilities are calculated as a set of composite probabilities using equation [**2**], as shown in Sheet 2, which gives spreadsheet functions that compute the respective probabilistic elements in the category probability set. Sheet 1 gives the computed probabilities for this example in the "Categorical Probabilities" section.

Finally, the consensus probabilities are computed using the simplification rule (equation [**1**]) in the "Consensus" section of Sheet 2. The results of the actual calculations are given in the "Consensus" section of Sheet 1. The probabilities are combined from top down, starting with the gender category and ending with race.

The results are more sensitive to demographics then they are to the preferences, as evidenced through experimentation with the spreadsheet recalcalculation facility.

SUMMARY

An admixture of methods has been given to structure a problem domain into categories and to compute categorical probabilities from structure elements and preference sets. The categorical probabilities are then combined using Dempster's rule of combination to obtain a composite assessment of the decision landscape. A demonstrative product allocation application is given. A sales organization must make an assessment of the products to stock and product analytics provides a methodology of formalizing the selection process. The methods can also be applied to product characteristics, as well as to products.

REFERENCES

1 Dempster, A.P. 1967, "Upper and Lower Probabilities Induced by a Multivalued Mapping," *The Annals of Statistics* 28:325-339.
2 Katzan, H. 1992, *Managing Uncertainty: A Pragmatic Approach*, New York: Van Nostrand Reinhold Co.
3 Katzan, H. 2006, "Consensus," *Proceedings of the Decision Science Institute National Conference*, San Antonio TX. (November 2006).
4 Katzan, H. 2008, "Categorical Analytics Based on Consensus Theory," *Journal of Business and Economics Research*, Volume 6, Number 8 (August, 2008).
5 Shafer, G. 1976, *A Mathematical Theory of Evidence*, Princeton, NJ: Princeton University Press
6 Theil, H. 1967, *Economics and Information Theory*, New York: American Elsevier Publishing Company, Inc.

	B	C	D	E	F	G	H	I	J	K	L	M	N
1	Product Consensus (A/B)				Categorical Probabilities (A/B)				Demographics			Product Preferences (A/B)	
2	0.58	0.42		Gender	0.58	0.42		Gender	Male	0.4		0.7	0.3
3									Female	0.6		0.5	0.5
4	0.520391517	0.479608483		Age	0.44	0.56		Age	<=25	0.4		0.2	0.8
5									26-55	0.4		0.5	0.5
6									>=56	0.2		0.8	0.2
7	0.480327332	0.519672668		Education	0.46	0.54		Education	None	0.4		0.2	0.8
8									Attended	0.2		0.5	0.5
9									Grad	0.4		0.7	0.3
10	0.530447127	0.469552873		Race	0.55	0.45		Race	B/C	0.1		0.2	0.8
11									Cauc	0.8		0.6	0.4
12									other	0.1		0.5	0.5

Sheet 1. Spreadsheet for the Product Allocation Application showing Demographics, Preferences, Categorical Probabilities, and the Product Consensus for products A and B.

	B	C	D	E	F	G
1	Product Consensus (A/B)				Categorical Probabilities (A/B)	
2	=F2	=G2		Gender	=K2*M2+K3*M3	=K2*N2+K3*N3
3						
4	=(B2*F4)/(B2*F4+C2*G4)	=(C2*G4)/(B2*F4+C2*G4)		Age	=K4*M4+K5*M5+K6*M6	=K4*N4+K5*N5+K6*N6
5						
6						
7	=(B4*F7)/(B4*F7+C4*G7)	=(C4*G7)/(B4*F7+C4*G7)		Education	=K7*M7+K8*M8+K9*M9	=K7*N7+K8*N8+K9*N9
8						
9						
10	=(B7*F10)/(B7*F10+C7*G10)	=(C7*G10)/(B7*F10+C7*G10)		Race	=K10*M10+K11*M11+K12*M12	=K10*N10+K11*N11+K12*N12

Sheet 2. Spreadsheet for the Product Allocation Application Giving Functions for the Calculations in Sheet 1.

END OF CHAPTER 9

10

STRUCTURAL ANALYTICS FOR DECISION MAKING UNDER UNCERTAINTY

INTRODUCTION

The effectiveness of unstructured decisions made under uncertainty directly involves two important concepts: the representation of the problem domain and the completeness of the solution space. A *category* is a means of representing the problem domain so that relevant structural information may be assessed and decisions can be made. A *frame of discernment* (Shafer [1976], Katzan [1992, 2006, 2008]) is a set of mutually exclusive and collectively exhaustive possibilities for the solution space.

Category

A category is a means of structuring a problem domain so that relevant information may be assessed and unstructured decisions can be made. The objective is to engage in a predictive modality in which one or more future events may be identified and analyzed.

Let $\mathbf{C}_i$ be one of the categories used to stratify the problem domain such that the collection

$$\mathbf{C} = \{\mathbf{C}_1, \mathbf{C}_2, \ldots, \mathbf{C}_n\}$$

represents a complete conceptualization of the dynamics under investigation and n is the number of categories.

Associated with each category is a set of probabilities representing an assessment of a future outcome based on its underlying categorical imperative. Thus, a category is a mechanism for isolating a single view of the problem under consideration. The ontological definition of a *category,* as a conceptual entity with no attributes in common with other categories, is adopted in this paper.

Frame of Discernment

A frame of discernment is a means of representing the possibilities under consideration, as in the following examples:

E = {McCain, Obama}
M = {Up, Unchanged, Down}

Clearly, the elements in a frame of discernment are in fact propositions that can be interpreted as events or states. Thus, if component s_i of system S over domain V were associated with the symbol "McCain," then that state is equivalent to the proposition, "The true value of V for component s_i is McCain," or in ordinary language, "s_i prefers McCain."

Accordingly, the set S of propositions S_i,

$$S = \{S_1, S_2, \ldots, S_n\}$$

commonly represent the collection of states of a system under analysis. Clearly, at an agreed upon point in time, one proposition is true and the others are false. If proposition S_i is true, then we can assert that

$$S_i = T \text{ and}$$
$$S_j = F, \text{ where } i \neq j$$

where T and F represent "true" and "false," respectively. The solution space contains the probabilities that each of the respective states is true.

Uncertainty

Prior to an agreed point in time (τ), we obviously do not know the state of the system under analysis or its components with any degree of certainty. The expectation that a part of the system will be in a particular state at time τ is denoted by a real number $p(S_i)$ associated with each of the propositions in the frame $S = \{S_i\}$, $i=1,2,\ldots,n$, such that

$$0 \leq p(S_i) \leq 1$$

and

$$\sum_{i=1}^{n} p(S_i) = 1$$

It follows that the function p can be extended to the lattice $\underline{S}$ of S such that

$$p(k) = \sum_{i=1}^{k} p(S_i) = 1$$

where k is an element of $\underline{S}$. This is simply the addition rule for mutually exclusive events.

Consensus theory is covered first, because it determines where we are going with the analysis. It is followed by a presentation of an approach to the structural analysis of categories. Finally, an election application is used as a demonstrative example that gives some insight into how the methods can be applied to other problems.

CONSENSUS THEORY

Consensus theory is a methodology for combining evidence based on Dempster-Shafer theory (Shafer [1976], Katzan [1992, 2006, 2008]) and the mathematical combination of evidence (Dempster [1967]). Dempster-Shafer theory has commanded a considerable amount of attention in the scientific and business communities, because it allows a knowledge source to assign a numerical measure to a proposition from a problem space, and provides a means for the measures accorded to independent knowledge sources to be combined. Dempster-Shafer theory is attractive because conflicting, as well as confirmatory, evidence from multiple sources may be combined.

The basis of Dempster-Shafer theory is the frame of discernment (Θ), introduced previously. Accordingly, a knowledge source may assign a numerical measure to a distinct element of Θ, which is equivalent to assigning a measure of belief to the corresponding proposition. In most cases, the numerical measure will be a basic probability assignment. A measure of belief may also be assigned to a subset of Θ or to Θ itself.

Support Functions

Consider a frame of discernment Θ and its power set denoted by 2^Θ. For example, given the frame:

$\Theta = \{a, b, c\}$

The power set is delineated as follows:

$$2^{\Theta} = \{\{a, b, c\}, \{a, b\}, \{a, c\}, \{b, c\}, \{a\}, \{b\}, \{c\}\}$$

In Dempster-Shafer theory, a knowledge source apportions a unit of belief to an element of 2Θ. This belief can be regarded as a mass committed to a proposition and represents a judgment as to the strength of the evidence supporting that proposition. When viewed in this manner, evidence focuses on the set corresponding to a proposition; this set is called a *focal set.*

The support for a focal set is a function m that maps an element of 2^{Θ}, denoted by A, onto the interval $[0,1]$. Given a frame of discernment Θ and function $m: 2^{\Theta} \rightarrow [0,1]$, a support function is defined as follows:

$m(\phi) = 0$, where ϕ is the null set
$0 \leq m(A) \leq 1$, and

$$\sum_{A \subset 2^{\Theta}} m(A) = 1$$

The support function m is called a *basic probability assignment,* which is assigned by the knowledge engineer or domain specialist.

A support function is called a *simple support function* if it reflects, at most, one focal set not equal to Θ. A simple support function assigns a measure of belief to the focal set A, as follows:

$m(A)>0$
$m(\Theta)=1-m(A)$
$m(B)=0$, for all $B \subset 2^{\Theta}$ and $B \neq A$

The simple support function for a focal set A assigns a portion of the total belief exactly to A and not to its subsets or supersets. The remainder of the belief is assigned to Θ. Because certainty function must add up to 1, $m(\Theta)=1\text{-}m(A)$.

It is possible that a body of knowledge or evidence supports more than one proposition, as in the following case. If

$$\Theta = \{a, b, c, d\}$$
$$A = \{a, b\}$$

and

$$B = \{a, c, d\}$$

then the evidence supports two focal sets, which in the example, are A and B. If $m(A)=0.5$ and $m(B)=0.3$, then $m(\Theta)=0.2$. A support function with more than one focal set is called a *separable support function*. Separable support functions are normally generated when simple support functions are combined.

The notion of combining simple support functions is a practical approach to the assessment of evidence. An analyst obtains information from a knowledge source, and it leads to an immediate conclusion – not with certainty, but with a certain level of belief. This is a normal straightforward means of handling human affairs and is precisely what people do. Then when additional information comes in, the various pieces of evidence are combined to obtain a composite picture of the situation.

Combination of Evidence

A method of combining evidence is known as Dempster's rule of combination (Dempster [1967]). Evidence would normally be combined when it is obtained from two different observations, each over the same frame of discernment. The combination rule computes a new support function reflecting the consensus of the combined evidence.

If m_1 and m_2 denote two support functions, then their combination is denoted by $m_1 \oplus m_2$ and is called their *orthogonal sum.* The combination $m_1 \oplus m_2$ is computed from m_1 and m_2 by considering all products of the form $m_1(X) \bullet m_2(Y)$, where X and Y range over the elements of Θ; $m_1(X) \bullet m_2(Y)$ is the set intersection of X and Y combined with the product of the corresponding probabilities.

For example, consider the frame of discernment

Θ = {healthy, tests, sick}

and views A and B, based on two different observation over the same frame:

A = {{healthy},0.6},{{tests},0.3},{{sick},0.1}}
B = {{healthy},0.4},{{tests},0.4},{{sick},0.2}}

The combination rule is depicted via the following tableau:

		A	
m_1	{healthy}	{tests}	{sick}
m_2	0.6	0.3	0.1
B			
{healthy}	{healthy}	{Ø}	{Ø}
0.4	0.24	0.12	0.04
{tests}	{Ø}	{tests}	{Ø}
0.4	0.24	0.12	0.04
{sick}	{Ø}	{Ø}	{sick}
0.2	0.12	0.06	0.02

The entries are then combined as follows:

$m_1 \oplus m_2(\{\text{healthy}\}) = 0.24$
$m_1 \oplus m_2(\{\text{tests}\}) = 0.12$
$m_1 \oplus m_2(\{\text{sick}\}) = 0.02$
$m_1 \oplus m_2(\{\varnothing\}) = 0.62$

Thus, for $A_i \cap B_j = \boldsymbol{A}$ and $m_1 \oplus m_2 = \boldsymbol{m}$, the combination rule is defined mathematically as:

$$\boldsymbol{m}(\boldsymbol{A}) = \sum_{A_i \cap B_j = \boldsymbol{A}} m_1(A_i) \bullet m_2(B_j) / (1 - \sum_{A_i \cap B_j = \varnothing} m_1(A_i) \bullet m_2(B_j))$$

The denominator reflects a normalization process to insure that the pooled values sum to 1. So, in this instance, the normalization process yields the combination

$$A \oplus B = \{\{\text{healthy}\},0.63\},\{\{\text{tests}\},0.32\},\{\{\text{sick}\},0.05\}\}$$

after normalization by dividing the combined assessment by (1-0.62) or 0.38. Because the problem is well-structured, the representation can be simplified as

$$A \oplus B = \{0.63,0.32,0.05\}$$

For views A=$\{A_1,A_2,\ldots,A_n\}$ and B=$\{B_1,B_2,\ldots,B_n\}$, the combination rule can be simplified as

$$A \oplus B = \{A_1 \times B_1/k, A_2 \times B_2/k, \ldots, A_n \times B_n/k\} \quad \textbf{[1]}$$

where $k = \sum_{i=1}^{n} A_i \times B_i$

We will refer to equation [**1**] as the *simplification rule.*

STRUCTURAL ANALYTICS

A problem domain is composed of categories, each of which is defined by a set of alternate structures. In a make-believe universe, for example, the category party could be defined as

party = {republican, democrat, independent}

based on a structural assessment, such as demographics. In this instance, the category party is one of many viewpoints of an underlying decision situation, which could be a vote in an election or on an important issue. We are going to argue that in many unstructured decision-making problems, the probabilistic outcome can be based on structural, rather than, preferential elements. What makes an unstructured decision so complex is that there are usually several categories "tugging at the decision maker." We are going to show how categorical assessments can be combined to form a composite assessment of a decision under consideration. Through the technique of structural analysis, we are going to assign probabilities to the elements of the frame of discernment from a given category, and then use consensus theory to combine the various probabilities. For example, a vote based on party could go one way and a vote based on gender could go another way. In reality, it would be a combination of the two views.

Structural Elements

Each category $\mathbf{C}_i$ is comprised of a set of structural elements $\mathbf{S}_i = \{S_{i1}, S_{i2}, S_{i3}, \ldots, S_{imi}\}$, where *mi* is the number of structural elements in category $\mathbf{C}_i$. In this analysis, we are going to stratify the election domain into eight categories as follows:

C = {party, gender, college, religion, attitude, age, region, race}, where

party = {rep, dem, indep},
gender = {male, female},

college = {none, attended, grad},
religion = {cath, prot, evan, jewish, other},
attitude = {conserve, liberal, neither},
age = {<=25, 26-55, >=56},
region = {N, S, M, W}, and
race = {B/C, cauc, other}

Each problem domain is represented by a set of categories.

Structural Probabilities

Accordingly, for each category $\mathbf{C}_i$ and its structure $\mathbf{S}_i$, the probability set would be expressed as:

$$\mathbf{P}_i = \{\text{prob}(S_{i1}), \text{prob}(S_{i2}), \ldots\}$$

For example, the probability set for category #1 (party), could be

$$\mathbf{P}_1 = \{0.4, 0.5, 0.1\}$$

representing rep, dem, and indep respectively.

Preference Set and Composite Probabilities

Each structural element has a preference set reflecting the probability that a voter in the respective domain will vote for an element of the solution space. In this example, the frame of discernment is {R,D}. In this case, a person in Party/Rep would vote for R with probability 0.7 and for D with probability 0.3.

Composite categorical probabilities for each element in the frame of discernment are computed by combining the structural probabilities and corresponding preference set as follows:

$$P(\Theta_{it}) = \sum_{j=1} ((\text{prob}(S_{ij}) \cdot \text{ps}(S_{ij})) \qquad [2]$$

where the index i runs through the categories and the index t runs through the alternatives in the frame of discernment, and $ps(S_{ij})$ reflects preference set preobabilities.

Categorical Probabilities

The composite probabilities represent a summation of the preference for each element of the frame of discernment for each category. The result is a set of independent categorical assessments of the problem domain from different viewpoints represented as probabilities computed using equation [**2**], and resulting in the following set:

$$
\begin{aligned}
\text{party} &= \{0.54, 0.46\} \\
\text{gender} &= \{0.58, 0.42\} \\
\text{college} &= \{0.46, 0.54\} \\
\text{religion} &= \{0.48, 0.52\} \\
\text{attitude} &= \{0.56, 0.44\} \\
\text{age} &= \{0.44, 0.56\} \\
\text{region} &= \{0.54, 0.46\} \\
\text{race} &= \{0.55, 0.45\}
\end{aligned}
$$

Sheet 1 gives the computed probabilities for this example in the "Categorical Probabilities" section.

Consensus

Finally, the categorical probabilities are combined using the equations for the simplification rule (equation [**1**]) shown in the "Consensus" section of Sheet 2. The results of the actual calculations are given in the "Consensus" section of Sheet 1. The probabilities are combined from top down, starting with the Party category and ending with Race. The results are more sensitive to demographics then they are to the preferences, as evidenced through experimentation with the spreadsheet recalc facility.

SUMMARY

An election example has been presented that structures a problem domain into categories and computes categorical probabilities from structural elements and preference sets. The categorical probabilities are then combined using Dempster's rule of combination to obtain a composite assessment of the election landscape.

REFERENCES

Dempster, A.P. (1967), Upper and Lower Probabilities Induced by a Multivalued Mapping, *The Annals of Statistics* 28:325-339.

Katzan, H. (1992), *Managing Uncertainty: A Pragmatic Approach*, New York: Van Nostrand Reinhold Co.

Katzan, H. (2006), Consensus, *Proceedings of the Decision Science Institute National Conference*, San Antonio TX. (November 2006).

Katzan, H. (2008), Categorical Analytics Based on Consensus Theory, *Journal of Business and Economics Research*, Volume 6, Number 8, August 2008.

Shafer, G. (1976), *A Mathematical Theory of Evidence*, Princeton, NJ: Princeton University Press

Theil, H. (1967), *Economics and Information Theory*, New York: American Elsevier Publishing Company, Inc.

	B	C	D	E	F	G	H	I	J	K	L	M	N
1	CONSENSUS (R/D)			Categorical Probabilities (R/D)				Demographics				Preferences (R/D)	
2	0.54	0.46		Party	0.54	0.46		Party	Rep	0.4		0.7	0.3
3									Dem	0.5		0.4	0.6
4									Indep	0.1		0.6	0.4
5	0.618483412	0.381516588		Gender	0.58	0.42		Gender	Male	0.4		0.7	0.3
6									Female	0.6		0.5	0.5
7	0.58	0.42		College	0.46	0.54		College	None	0.4		0.2	0.8
8									Attended	0.2		0.5	0.5
9									Grad	0.4		0.7	0.3
10	0.560386473	0.439613527		Religion	0.48	0.52		Religion	Cath	0.3		0.2	0.8
11									Prot	0.3		0.6	0.4
12									Evan	0.2		0.6	0.4
13									Jewish	0.1		0.7	0.3
14									Other	0.1		0.5	0.5
15	0.618666667	0.381333333		Attitude	0.56	0.44		Attitude	Conserv	0.3		0.8	0.2
16									Liberal	0.3		0.4	0.6
17									Neither	0.4		0.5	0.5
18	0.560386473	0.439613527		Age	0.44	0.56		Age	<=25	0.4		0.2	0.8
19									26-55	0.4		0.5	0.5
20									>=56	0.2		0.8	0.2
21	0.599425837	0.400574163		Region	0.54	0.46		Region	N	0.3		0.4	0.6
22									S	0.2		0.4	0.6
23									M	0.2		0.8	0.2
24									W	0.3		0.6	0.4
25	0.646512413	0.353487587		Race	0.55	0.45		Race	B/C	0.1		0.2	0.8
26									Cauc	0.8		0.6	0.4
27									other	0.1		0.5	0.5

Sheet 1. Spreadsheet for the Election Application showing Demographics, Preferences, Categorical Probabilities, and the Consensus for the Election Application.

	B	C	D	E	F	G
1	CONSENSUS (R/D)				Categorical Probabilities (R/D)	
2	=F2	=G2		Party	=K2*M2+K3*M3+K4*M4	=K2*N2+K3*N3+K4*N4
3						
4						
5	=(B2*F5)/(B2*F5+C2*G5)	=(C2*G5)/(B2*F5+C2*G5)		Gender	=K5*M5+K6*M6	=K5*N5+K6*N6
6						
7	=(B5*F7)/(B5*F7+C5*G7)	=(C5*G7)/(B5*F7+C5*G7)		College	=K7*M7+K8*M8+K9*M9	=K7*N7+K8*N8+K9*N9
8						
9						
10	=(B7*F10)/(B7*F10+C7*G10)	=(C7*G10)/(B7*F10+C7*G10)		Religion	=K10*M10+K11*M11+K12*M12+K13*M13+K14*M14	=K10*N10+K11*N11+K12*N12+K13*N13+K14*N14
11						
12						
13						
14						
15	=(B10*F15)/(B10*F15+C10*G15)	=(C10*G15)/(B10*F15+C10*G15)		Attitude	=K15*M15+K16*M16+K17*M17	=K15*N15+K16*N16+K17*N17
16						
17						
18	=(B15*F18)/(B15*F18+C15*G18)	=(C15*G18)/(B15*F18+C15*G18)		Age	=K18*M18+K19*M19+K20*M20	=K18*N18+K19*N19+K20*N20
19						
20						
21	=(B18*F21)/(B18*F21+C18*G21)	=(C18*G21)/(B18*F21+C18*G21)		Region	=K21*M21+K22*M22+K23*M23+K24*M24	=K21*N21+K22*N22+K23*N23+K24*N24
22						
23						
24						
25	=(B21*F25)/(B21*F25+C21*G25)	=(C21*G25)/(B21*F25+C21*G25)		Race	=K25*M25+K26*M26+K27*M27	=K25*N25+K26*N26+K27*N27
26						
27						

Sheet 2. Spreadsheet for the Election Application Giving Functions for the Calculations in Sheet 1.

END OF CHAPTER 10

APPENDIX

A BRIEF INTRODUCTION

The Dempster-Shafer theory of evidence has commanded a considerable amount of attention in the scientific and business communities, because it allows a knowledge source to assign a numerical measure of belief to a proposition from a problem space and provides a means for the beliefs accorded independent knowledge sources to be combined. Dempster-Shafer is attractive because conflicting, as well as confirmatory, evidence from multiple agents may be combined.

The basis of Dempster-Shafer theory is a set of distinct and exhaustive possibilities called a *frame of discernment.* By convention, the frame of discernment is demoted by Θ, such as in the following example:

$\Theta = \{\text{healthy, tests, sick}\}$

Another way of viewing the concept is as follows: if *q* is a variable, then Θ represents the set of all possible values of *q*. The elements of a frame of discernment (usually called simply a "frame") correspond to a proposition, such as:

"The true value of *q* is healthy."

Clearly, a knowledge source may assign a numerical measure of belief to a distinct element of Θ, which is equivalent to assigning a measure of belief to the corresponding proposition.

A measure of belief may also be assigned to a subset of Θ or to Θ itself. If A is a subset of Θ, such as:

A = {test, sick}

Then assigning a belief function to A is equivalent to assigning that belief to the disjunctive proposition:

"The true value of *q* is *tests* or *sick*."

The capability of dealing with uncertain evidence that is restricted to a subset of a frame and at then same time being able to associate a measure of uncertainty to it gives Dempster-Shafer theory much of its intuitive and practical appeal.

Other than the ability to combine evidence, another important aspect of Dempster-Shafer theory is the manner in which complementary evidence is handled. In traditional probability theory, assigning a belief to a proposition implicitly commits the remainder of one's belief to its complement. For example, given the frame:

$\Theta = \{a, b, c, d\}$

subset

$A = \{a, c\}$

and belief

$\text{Bel}(A) = \text{Bel}(\{a,c\}) = s$

then,

$\text{Bel}(A^c) = \text{Bel}(\{b, d\}) = 1\text{-}s$

In Dempster-Shafer theory, the second requirement is relaxed. Thus, one can assign a belief of s to a subset A of Θ and leave the remainder of his or her belief uncommitted, as follows:

$$\text{Bel}(A) = s$$
$$\text{Bel}(\Theta) = \text{Bel}(\{a, b, c, d\}) = 1\text{-}s$$

This is the same as saying, "I commit a belief of s to A, and I am unsure of the remainder of my belief, which I am assigned to the frame." It is important to note that committing an amount of belief to the frame of discernment represent a requisite lack of knowledge.

From the viewpoint of subjective probability, the freedom inherent in Dempster-Shafer theory in intuitively appealing. On the other hand, Bayesian probability continues to be relevant to the subject of "chance functions" in classical statistics and probability theory.

Frame of Discernment

The frame of discernment is the set of possibilities under consideration. Choosing an appropriate formulation for a problem is crucial, because the frame does *not* exist independently of the decision maker's knowledge. Representation is the creative aspect of knowledge engineering. In Dempster-Shafer theory, the elements of the frame acquire their meaning from what the knowledge engineer and decision maker know or think they know.

By convention, the frame of discernment is demoted by Θ, a covered above. The set of all the subsets of Θ is denoted by 2^{Θ}. For example, given the frame:

$$\Theta = \{a, b, c\}$$

the set of subsets of Θ (i.e., the power set of Θ) is delineated as follows:

$$2^{\Theta} = \{\{a, b, c\}, \\ \{a, b\}, \{a, c\}, \{b, c\} \\ \{a\}, \{b\}, \{c\}\}$$

Because the null set yields no information, it is commonly omitted when delineating the contents of 2^{Θ}.

Simple Support Functions

In Dempster-Shafer theory, a knowledge source apportions a unit of belief to an element of 2^{Θ}. This belief can be regarded as a mass committed to a proposition and represents a judgment as to the strength of the evidence supporting that proposition. When viewed in this manner, evidence focuses on the set corresponding to a proposition; the set is called a *focal set.*

The support for a focal set is a function m that maps an element of 2^{Θ}, denoted by A, onto the interval $[0, 1]$. Given a frame of discernment Θ and the function:

$$m: 2^{\Theta} \rightarrow [0, 1]$$

a *support function* is defined as follows:

$$m(\Phi) = 0, \text{ where } \Phi \text{ is the null set}$$
$$0 \leq m(A) \leq 1, \text{ and}$$

$$\sum_{A \subset 2^{\Theta}} \mathrm{m}(A) = 1$$

Shafer calls the support function m a *basic probability assignment,* which is "assigned" to a proposition by a knowledge engineer or decision maker.

A support function is called a *simple support function* if it has at most one focal set not equal to Θ. A simple support function assigns a measure of belief to the focal set A, as follows:

$$m(A) > 0$$

$$m(\Theta) = 1 - m(A), \text{ and}$$

$$m(B) = 0, \text{ for all } B \subset 2^{\Theta} \text{ and } B \neq \mathbf{A}$$

The simple support function for a focal set A assigns a portion of the total belief exactly to A, and not to its subsets or supersets. The remainder of the belief is assigned to Θ. Because the certainty functions must add up to 1, then $m(\Theta)=1-m(A)$.

Consider, for example, the focal set

$$G = \{\text{mod, low}\}$$

Corresponding to the proposition, "The true value of G is *mod* or *low*," over the frame of discernment

$$\Theta = \{\text{hi, mod, low, unc}\}$$

and the support function k. If k assigns a measure of belief to G as follows:

$$k(G) = 0.35$$

$$k(\Theta) = 1 - k(G) = 0.65$$

then k is a simple support function.

If the focal set is the entire frame of discernment, which represents total ignorance, then we have a vacuous belief.

On to Dempster-Shafer Theory

Next comes the separable support function, belief functions, and the combination of evidence. Those and other topics are covered in the text.

Source: Katzan, H., *Managing Uncertainty: A Pragmatic Approach*, New York: Van Nostrand Reinhold, 1992. Pages 31-34.

INDEX

A

Action item 11–13, 23
Advancement of learning 47, 59
Advisor model 15–18
AI 35–40, 57, 81, 121, 138
AI conference 37
AI technology 39
Analytic knowledge 48, 93
Antecedent 8, 9, 22
Artificial intelligence 13, 23, 34–40, 159
Assignment of belief 31
Attribute 16, 18, 44, 46–51, 62, 89–90, 92–94, 96, 108–109, 115, 132

B

Bacon 47, 59
Basic probability assignment 28, 31, 78–79, 83, 117–118, 134–135, 148
Bayesian 3, 20, 147
Belief 3, 7, 11, 21, 23, 25–29, 31–33, 43, 45–47, 50–56, 67, 78–80, 109–110, 117–119, 134–136, 145–150
Belief structures 3, 11, 43, 50–51, 56
Biological metaphor 64
Biometrics 59, 100, 101
Bourne identity 49, 93

C

Categorical 38, 59, 73, 114–115, 123–124, 126–129, 132, 139–143
Category 49, 61–70, 93, 97, 108, 113–116, 122–128, 131–132, 134, 139–142
Cognitive science 36, 38–39
Combination of evidence 29, 33, 51, 56, 80, 85, 106, 108–109, 111, 117, 120, 134, 136, 150
Compatibility relation 8–10, 13, 51, 53–55
Compliance 60, 106, 112
Composite probabilities 126–127, 140–141
Consensus theory 51, 59, 73, 111, 114–115, 117, 123, 128, 134, 139, 142
Consequent 8–9, 22
Cybercrimes 87–88

D

Data profiling 100, 102–104
Decision script 8–10, 12, 23
Decision support system 8, 11, 13, 20
Democratization 64–66, 113, 115
Demographics 66, 113, 114, 123–125, 127, 129, 139, 141, 143
Dempster x, 5, 10, 12, 21–22, 25–29, 33, 35, 51, 55–59, 78, 80,

83–85, 110, 117–118, 120, 128, 134–136, 142, 145–148, 150
Dempster-Shafer viii, 12, 21, 25–28, 33, 35, 51, 78, 85, 117–118, 134–135, 145–148, 150
Derivative knowledge 48, 93
Description viii, 16–17, 48–50, 66, 88, 92, 94, 100
Direct knowledge 46, 48, 92
Dispositional belief 46
Domain 1–4, 6–7, 9, 17, 19, 21–23, 46, 49–50, 52, 61–64, 66, 69, 74, 88, 92–94, 100, 103, 111, 113, 115–116, 118, 122, 124, 126, 128, 131–132, 135, 139, 140–142
DSS 8–13

E

Elementary operation 63
Entropy 76–77, 83–85, 122
Epistemic belief 45
Evidence 3, 5–7, 9, 11, 13, 18–19, 21–23, 25–26, 28–31, 33–34, 45, 47–48, 51–57, 60, 78–80, 83, 85, 92–93, 106, 108–109, 111, 117–120, 127–128, 134–136, 142, 145–146, 148, 150
Evolution 2, 62, 64, 69–70
Evolutionary dynamics 61–62, 71–72
Evolutionary service element 62–63
Expert opinion 20, 82, 84–85, 121–122
Explicit belief 45

F

Fast food 62–63
Fitness 64–65, 67–71, 103
Frame 3–7, 9–12, 25–28, 30–33, 52–55, 57, 62, 64, 66, 68–69, 74–75, 78–81, 83–84, 113, 116–118, 120–123, 125–127, 131–137, 139–141, 145–150
Frame of discernment 3–4, 5, 7, 10–11, 25, 27–28, 30, 32–33, 52, 57, 62, 64, 66, 69, 74, 78–80, 83–84, 113, 116–118, 120–122, 123, 125–127, 131–132, 134–137, 139–141, 145, 147–150
Functionality 1–2
Fusing of evidence 5

G

Governance 60, 106, 112

H

Human-machine cognition system 38, 40

I

Identity 43–44, 46–52, 56, 59, 60, 87–95, 98, 100–103, 105–108, 111–112, 157
Identity analytics 43, 52, 108, 112
Identity theory 46, 49, 89, 92–94, 106, 108, 111
Indicator 2–6, 12–13, 18–21
Indirect knowledge 48, 92
Indiscernible 49, 59, 93–94, 112
Information domain 1

Information theory 75–76, 85, 128, 142
Input mapping 3–6, 10, 13
Input space 3–5, 7–10, 13
Intelligence 13, 23, 34–40, 98, 159
Intelligence records 98

K

Kant 48, 59
Katzan viii, 13, 23, 34, 46, 51, 58–59, 61, 63, 65, 71, 92, 95, 97, 108, 110, 112–114, 117, 128, 131, 134, 142, 150, 155, 157
Knowledge 2, 7, 21, 22, 25– 27, 29, 31, 38, 46–47, 48, 51–53, 55–56, 76, 78–80, 92–93, 107–110, 117–119, 125, 134–136, 145, 147–148

L

Logic theorist 37
Lusch 61, 71

M

McCarthy 37
Medical 4, 22, 62–63, 66, 103, 108
Monetization 65–66
Mutation 61, 63, 68, 70–71
Mutation matrix 70

N

Newell 37
Nowak 61, 66, 68, 71
Numerical identity 49, 93

O

Olfati-Saber 66, 68–69, 71, 72
Ontological 88, 115, 132
Output mapping 9
Output space 4, 7–13

P

Power 27, 47, 52, 76, 78, 97, 118, 134, 135, 148
Pragmatic belief 45–46
Preference set 125, 126–128, 140–142
Prior belief 55
Privacy 43–47, 51, 59–60, 87–92, 95–100, 102–106, 111–112, 157
Privacy and freedom 95
Probability assignment 28, 31, 78, 79, 83–84, 117–118, 122, 134–135, 148
Problem domain 2–3, 7, 17, 21–22, 61, 74, 113, 115, 122, 124, 126, 128, 131–132, 139–142

Q

Qualitative identity 49, 93

R

Replication 61, 63–64, 67–68, 71
Risk management 106

S

Separable support function 5, 29, 53, 80, 119, 136, 150
Service demographics 66
Service dynamics 64–65, 68
Service evolution 62, 69
Service kinematics 67
Service provisioning 61, 65, 71
Service system 62–63, 67, 72

Shannon 37
Simon 37
Simon and Newell 37
Simple support function 5, 28–29, 53, 79–80, 119, 135–136, 148–149
Solution domain 7, 22–23
Specificity 1–2
Spohrer 61, 72
Statistical records 98
Structural 113–114, 116, 122–127, 131, 134, 139–140, 142
Subject ix, 22, 37–39, 43–51, 53, 56, 65, 71, 88–100, 104–105, 107–109, 111, 147
Support function 5, 28–30, 52–53, 57, 78–80, 83–84, 118–120, 122, 134–137, 148–150
Synthetic knowledge 48, 93
System 1–3, 7–13, 15–18, 20, 27, 36–40, 43–46, 49, 52, 59, 62–63, 67, 71–77, 82–85, 88–92, 94, 98, 102, 104, 108, 111, 116, 121, 132–133, 157, 158

T

Temporal identity 49–50, 93–94

U

Uncertainty vii–viii, xi, 1–3, 12–13, 15, 17–20, 22–23, 26, 31, 34, 41, 48, 59, 75, 85, 93, 113, 116, 128, 131, 133, 142, 146, 150, 158–159

V

Value 1, 3–9, 12–13, 17, 19, 21, 25–26, 28, 30–33, 39, 46–50, 52, 56, 58, 61, 65, 67, 69–70, 74, 76–77, 82, 90, 92–93, 96, 110, 114, 116, 121, 125, 132, 138, 145–146, 149
Vargo 61, 66, 71–72
Video surveillance 100, 102

W

Wine advisor 16, 18
Workplace monitoring 100, 102

ABOUT THE AUTHOR

Harry Katzan, Jr. is a professor who has written several books and many papers on computers and service, in addition to some novels. He has been a advisor to the executive board of a major bank and a general consultant on various disciplines. He and his wife have lived in Switzerland where he was a banking consultant and a visiting professor. He is an avid runner and has completed 94 marathons including Boston 13 times and New York 14 times. He holds bachelors, masters, and doctorate degrees.

BOOKS BY HARRY KATZAN, JR.

COMPUTERS AND INFORMATION SYSTEMS

Advanced Programming
APL Programming and Computer Techniques
APL Users Guide
Computer Organization and the System/370
A PL/I Approach to Programming Languages
Introduction to Programming Languages
Operating Systems
Information Technology
Computer Data Security
Introduction to Computer Science
Computer Systems Organization and Programming
Computer Data Management and Database Technology
Systems Design and Documentation
Microprogramming Primer
The IBM 5100 Portable Computer
Fortran 77
The Standard Data Encryption Algorithm
Introduction to Distributed Data Processing
Distributed Information Systems
Invitation to Pascal
Invitation to Forth
Microcomputer Graphics and Programming Techniques
Invitation to Ada
Invitation to Ada and Ada Reference Manual
Invitation to Mapper
Operating Systems (2nd Edition)
Local Area Networks
Invitation to MVS (with D. Tharayil)
Privacy, Identity, and Cloud Computing

BUSINESS AND MANAGEMENT

Multinational Computer Systems
Office Automation
Management Support Systems
A Manager's Guide to Productivity, Quality Circles, and Industrial Robots
Quality Circle Management
Service and Advanced Technology

RESEARCH

Managing Uncertainty

SERVICE SCIENCE

A Manager's Guide to Service Science
Foundations of Service Science
Service Science
Introduction to Service
Service Concepts for Management
A Collection of Service Essays
Hospitality and Service

NOVELS

The Mysterious Case of the Royal Baby
The Curious Case of the Royal Marriage
The Auspicious Case of the General and the Royal Family
A Case of Espionage
Shelter in Place
The Virus
The Pandemic
Life is Good

The Vaccine
A Tale of Discovery
The Terrorist Plot
An Untimely Situation
The Final Escape

LITTLE BOOKS

The Little Book of Artificial Intelligence
The Little Book of Service Management
The Little Book of Cybersecurity
The Little Book of Cloud Computing
The Little Book of Managing Uncertainty

END OF BOOK – THANK YOU FOR READING IT

Printed in the United States
by Baker & Taylor Publisher Services